I0490028

Paola Morgese

HANDBOOK FOR
SUSTAINABLE PROJECTS

Global sustainability and project management

HANDBOOK FOR SUSTAINABLE PROJECTS
Global sustainability and project management

ISBN: 9781499397406

Copyright © Paola Morgese, 2014. All rights reserved.
All rights are reserved, including the total or partial, also automatic, translation in other languages. No part of this book can be reproduced or transmitted in any form or by any means, electronic, manual, photocopying, recording, or by any information storage and retrieval system, without prior written permission of the author. This book cannot be sold or lent, or hired-out without the written authorization of the author.

Photography on the cover page: Matese Park, Italy (© Paola Morgese, 2013).

To Nina

Table of contents

List of Figures

List of Tables

Foreword

This book comes at the end of my three years long experience as the first knowledge management leader of the Global Sustainability Community of Practice of the PMI® (Project Management Institute). I felt the need for further discussion in the field of the global (environmental, social, and economic) sustainability and of the management of projects, both of them my strong passions.

During this period, I have had the opportunity to interact with researchers, practitioners, students, writers, experts and beginners of different ages and from different parts of the world, both virtually and in person on the evolving and intricate issues related to sustainability. What is common among us is the passion for a new, fair, ethic, comprehensive, and long term vision of the projects.

Moreover, this book forms a summary of my work experiences, my, also international, seminars, webinars, blog posts, articles, and, university and non, lessons about the global sustainability in the management of the projects, and on the sustainable development.

In this book I'll try to give some practical approach to project management in an organized and structured manner, so that this passion would turn into positive, and beneficial for everyone, products, results, and services.

The book is intended for readers who already have an understanding and knowledge, even if minimum, in the field of project management and sustainability, and refers to the 2013 ISO 21500 standard "Guidance on Project Management".

Only some projects are sustainable by nature, but all projects can be managed sustainably.

Naples (Italy), Summer 2013

Paola Morgese

Introduction

The handbook details an organized and structured methodology for managing projects sustainably, and includes the following subjects:

- General definitions;
- Environmental sustainability;
- Social sustainability;
- Economic sustainability;
- Business environmental sustainability;
- Types of sustainable projects;
- Difference between investment and project;
- How to manage a project sustainably;
- Glossary of terms used in sustainability.

1. Definitions

1.1 Global sustainability and sustainable development

There are several definitions of sustainable development and of global sustainability. Here some of them are listed.

The first and most known definition of sustainable development, "*development that meets the needs of the present without compromising the ability of future generations to meet their own needs*", is written in the 1987 Brundtland Report of the World Commission for Environment and Development. The Brundtland Report highlighted the necessity to give priority to the needs of poor countries over the needs of rich countries.

The definition of global sustainability of the PMI® Project Management Global Sustainability Community of Practice is written from the first business plan dated 2009: "*global sustainability is the attainment of enduring economic, social, and environmental well-being of all elements of society*".

Another very well known definition is that one of John Elkington dated 1997, "*the triple bottom line*", and refers to the concept of *"people, planet, profit". "It is not possible to achieve a desired level of ecologic or social or economic sustainability (separately), without achieving at least a basic level of all three forms of sustainability, simultaneously".*

According to Gareis et al. (2013), it was the German Hans Carl von Carlowitz, at the beginning of the eighteenth century, who introduced for the first time the concept of sustainable development in his book on forestry.

1.2 Project

The definition of project from the dictionary can be separated into:
- Project: an ordered and detailed plan of work to execute something;
- Design: a set of calculations, drawings, and documents needed to define unequivocally the idea based on which any construction (product or good) can be created.

The definition of project from "A Guide to the Project Management Body of Knowledge" or PMBOK® Guide of the Project Management Institute (PMI®) (hereinafter "PMP, PMBOK and PMI are registered marks of the Project Management Institute, Inc. ") is:

"Temporary endeavor undertaken to create a unique product, service, or result".

Therefore, a project is:
- Temporary: each project has its well defined beginning and end;
- Unique: each project has its own characteristics and is different from the other projects.

2. Global sustainability

2.1 Global sustainability

Environmental sustainability, social sustainability, and economic sustainability are strongly connected with each other and influence each other.

2.2 Environmental sustainability

By way of, non-exhaustive, example, some characteristic items of the environmental sustainability are schematically described below.

About the management of waste, the production of waste should be reduced or, at least, less hazardous waste and more non hazardous waste should be produced and the possible alternatives for their treatment and disposal should be evaluated, choosing the most sustainable ones, preferring for instance composting to disposing of in landfill or to incineration. Reuse and recycling of the produced waste should be increased.

A significant reference for water can be extracted from Gareis et al. (2013) and from Vienna Hospital Association (2010). Among the 31 quality criteria required for planning and construction, the criteria 4 of the "Sustainability Charter" refers to the use of drinking water and of rain-water. Drinking water demand should be reduced to the minimum indispensable quantity and its use reduced with the installation of specific devices. Drinkable water should not be used for facilities, for instance washing or watering. Rain-water should be drained (also from the roofs) and let slowly naturally percolate in non paved areas, to promote the growth of plants and animals, or reused, instead of conveyed in the sewer.

While choosing the raw materials, the use of by-product raw materials and of reusable or recycled materials should be measured and preferred to the use of non renewable materials such as minerals, carbon, oil and its by-products, natural gas, and metals.

Devices and technologies to economize on power consumption should be used. Moreover, the use of energy from renewable sources, such as sun, wind, tides, hydroelectric, biomass, geothermal resources, (non alimentary) biofuels, and hydrogen, should be measured and preferred to the use of power from non renewable sources such as carbon, natural gas, oil and its by-products, and nuclear.

Emissions in atmosphere of the following substances should be measured and limited (GRI, 2011)(EPA, 2010) (EPA, 2011):
- Greenhouse gases, considered also among the main causes of the climate change: carbon dioxide CO_2, methane CH_4, nitrous oxide N_2O, hydrofluorocarbons HFCs, perfluorocarbons PFCs, and sulphur hexafluoride SF_6;
- Main ozone depleting substances: chlorofluorocarbons CFCs, hydrochlorofluorocarbons HCFCs, and halogens.

It is intended to highlight how it is important to make always measures, introducing for instance some partial meters to gauge the consumption of water and of energy, to evaluate correctly the actions made in the field of the environmental sustainability.

Other interventions could concern for instance the protection of biodiversity, or transportation, and depend anyway on the specific objectives decided in the particular project.

Some references in the environmental field are the ISO 14000 and the Kyoto protocol, adopted in 1997 and come into force in 2005, to control the gaseous emissions in atmosphere.

See also the glossary at the end of the book.

2.3 Social sustainability

Social sustainability is connected with (GRI, 2011) the respect for the:
- Human rights;
- Rights of the society;
- Rights of the workers;
- Rights of the consumers.

With regard to the human rights, the main reference is the "Universal Declaration of Human Rights" of the United Nations dated 1948, about civil, political, economic, social and cultural rights. The declaration is against discriminations, child labor, forced or compulsory labor (slavery) and is in favor of freedom of association, rights of ethnic minorities, access to medical care, to education, and to culture.

Human rights concern the accessibility for people suffering from disabilities to the use and fruition of products, services or results of projects.

The rights of the society concern the interactions with the market (connection with the economic sustainability), the institutions, and the local communities. They are in favor of legality, fairness, and transparency. They support wide public consultation and the needs of the local communities, such as the social and economic development, the opposition to pollution and to the exploitation of the natural resources (connection with the environmental sustainability), and the protection of public health. They are against corruption and monopoly.

With corruption it is meant bribery, fraud, extortion, collusion, conflict of interest, and money laundering. Corruption creates poverty (connection with the economic sustainability), endangers the environment (connection with the environmental sustainability), affects democracy and human rights, threatens legality, and misallocates investments (connection with the economic sustainability). Main reference in this field is the 2003 "United Nations Convention Against Corruption".

About the interactions with institutions, the rights of the society are against pressures for political purposes, the influence on the governmental policies with organized and coordinated actions, persuading or influencing people holding a public office to promulgate laws or to make political decisions, political contributions and funds aimed to give votes in exchange for donations, loans, publicity, use or donation of goods (for instance cars, boats, houses), hiring, and employment.

About the interactions with the market, they are in favor of a healthy competition among firms, promoting economic efficiency and sustainable growth, in opposition to monopoly and unfair competition towards competitors to fix prices, to coordinate bids, to create market and production restrictions, to impose geographic shares, to allot clients and suppliers (connection with economic sustainability).

The rights of the workers are based above all on the 1998 ILO (International Labour Organization) "Declaration on the Fundamental Principles and Rights at Work", and promote economic growth and equity.

They concern the existence of collective bargaining agreements, ensuring freedom of association and organization stability; an indiscriminate and balanced distribution between employees and contractors, between full time workers and part time workers, between temporary employees and permanent employees, between women and men, and a fair rate of turn over, positively affecting the human and intellectual capital and the productive capacity of the company.

They include a wide consultation, information and involvement of the workers and of their representatives, also for issues concerning reorganization, merger, expansion, sale, closure, opening.

They promote the prevention of accidents, injuries, professional diseases, deaths, and regular exposure to hazardous chemical substances. They include the involvement of the labor unions, the promotion of the culture of health and safety in the workplace, the assistance, the advice and the prevention, the development of specific campaigns of information and training for the workers, for their families, and for the whole community for particular risks (for instance the asbestos risk).

The rights of the consumers concern quality, health, safety, labelling, packaging, maintenance, end of life destination (for instance disposal or recycling and reuse), privacy, customer satisfaction, impacts in the use of the product or service (for instance home appliances with low water and power consumptions), non invasive and non annoying management of promotions, and the management of complaints.

The rights of the consumers can have remarkable influence on the choice of materials and on the production cycle, and should be considered in the long term.

Some references in the social field are the ISO 26000, Guidance on Social Responsibility, and the SAI (Social Accountability International) 8000.

See also the glossary at the end of the book.

2.4 Economic sustainability

To understand what the economic sustainability refers to, it is necessary to make a first distinction between economy, that is the efficient and rational use of resources in the production of goods and services, and finance, that is the mere management of the money.

The economic analysis refers to the community, from where it links with the social aspects, and differs from the financial analysis, which on the contrary pertains to the single private operator, who undertakes it (Forte, 1977).

The economic sustainability of the project concerns (Gareis et al., 2013) an efficient management of the resources of the project, the resolution of the project discontinuities, the management of the complexity, of the dynamics and of the relations of the project contexts, and the optimization of the economic results of the investment started with the project.

The economic sustainability of the project includes (Silvius et al., 2012) the guarantee of a profitable investment accompanied by agility and flexibility.

The business economic sustainability (GRI, 2011) concerns:
  − Economic value generated;

- Economic value distributed;
- Economic value retained (generated minus distributed).

The economic value generated is formed mainly by the company revenues. The economic value distributed concerns the wealth distributed to suppliers, to employees and contractors, to investors and shareholders, to public agencies, and to the community.

The specific percentages through which the economic value is distributed are also important, in order that the economic well-being of the workers, the support to the local working market, the positive indirect economic impacts, that are additional positive impacts caused by the circulation of the money through the local and regional economy, strictly connected with the sustainable development, and generally the social and economic improvement, for instance with the economic development in depressed areas, the diffusion of information technologies, and the local general improvement of the social, economic, professional and cultural conditions are ensured.

References in the economic field are the International Accounting Standards Board (IASB) and the International Financial Reporting Standards (IFRS).

See also the glossary at the end of the book and the connections with the social sustainability, described at the previous section.

2.5 Business environmental sustainability

In the last twenty years also the organizations began to apply the principles of the sustainable development to their production cycle, to measure performances and results in the social and environmental, other than in the economic, field, and to develop not only economic, but also social and environmental reports and balance sheets.

Putting down in black and white the numbers of the environmental sustainability and starting to make comparisons, positive surprises emerged comparing business activities managed traditionally and the same activities managed differently according to the environmental sustainability criteria. Initially, only the environmental risks have been taken into account, now on the contrary also opportunities are emerging.

The transformation of qualitative concepts into measurable and measured quantitative entities has revealed positive business results. The environmental sustainability made by definitions, notions, and theory about waste management, power saving, protection of resources, use of renewable energy, selection of suppliers, study of the entire life cycle of the product from raw materials to the final destination after use, all calculations considered, has produced improvements in the field of safety and environment, economic savings, solution of technical issues, solution of commercial issues, solution of management issues, and other tangible benefits for organizations.

The positive result refers not only to large organizations, but also to small and medium companies. Just the publication of a periodical environmental sustainability report turns into a good reputation for the firm and into an increase in the numbers of clients, more and more sensitive about the issue. An organization contributing to the continuous improvement of the environmental conditions, measuring, communicating, and accountable for the company results in this sector, automatically provides the image of a reliable, transparent, and responsible firm under all points of view.

If, previously, comparing the results attained in the field of the environmental sustainability with the parameters of law, regulations and norms was enough to avoid penalties and legal actions, now the goals to be attained are established with the organization policies and strategies themselves in a wider extent and serve both for an assessment of one's own skills and engagement, and for a comparison in the same field with other competing firms.

Business environmental sustainability has turned from the need to comply with laws and regulations in the sector to a market need and, as such, it has become integral part of the organization balance sheet.

The organization environmental sustainability report (GRI, 2011) considers first of all the parameters to be evaluated as required by the respect for laws and regulations, and extends furthermore to all of those factors, that have a significant economic and financial impact on the firm itself, with reference to the definition of sustainability of the 1987 Brundtland Report, fixing a range of priorities and considering the company with its connections with the places, where it is located. For instance, the polluting load of a firm should not be considered as an absolute value, but in relation to the characteristics of the local ecosystem. Or, both the parameters affecting the environment in the short term, and those ones affecting it in the long term, for instance persistent or bioaccumulable pollutants, and pollutants causing irreversible impacts should be considered.

Summarizing, some of the goals that a company can intend to attain in the environmental field, for itself, for its suppliers, and for its customers, are the use of clean and renewable energy, the reduction of waste, the appropriate collection, storage, and disposal of the remaining waste, the reuse, the recycling, the energy saving, the water reuse, the good quality of the work environment, the wise use of natural resources, avoiding pollution, protecting air, water and soil, and the knowledge and improvement of the natural environment.

All of what written till here can be numerically measured and, as such, elaborated, interpreted and compared while time elapses.

Some numeric evaluations of the environmental sustainability of a firm are already contained in its documents, reports, and archives, other ones can be extrapolated from them, and other ones more need specific measures, with the installation, for instance, of specific meters.

The data already available are all those ones referring to violations of laws and regulations that caused penalties, legal costs, expenses for damage compensations, expertise, and consultations, costs for remediation of contaminated sites, and to the connected consequences and repercussions, such as drop in sales, loss in reputation, production interruptions, and suspension in the provision of services.

Other numbers immediately available are those ones regarding the expenses that the company made for investments in the environmental field, for instance renovation and compliance of plants and equipment, technological updating, employee training, ISO 14000 certification, research and innovation in the sector, expenses for waste disposal, expenses for the treatment of wastewater and of gaseous emissions in atmosphere, and connected costs of operation and maintenance.

Further costs already in archive are those ones relating to environmental accidents, accidental leaks and discharge of more or less polluting substances, types of those leaks, and connected impacts on air, water and soil in the short and long term.

The next data, which can be extrapolated from the previous ones or to be specifically gauged, are the data that can help to review and to optimize the business production cycle and the habits of the organization, to reduce squandering, to save money, to increase the value of products and services, and to generally advance a firm in comparison to competitors, not only in the environmental sector.

For instance, the environmental costs connected with transportation, with reference also to the raw materials, to the final products, and to the company employees, could be calculated. All the activities connected with transportation imply a consumption of energy, and so then a cost. Other impacts of transportation refer to noise, polluting emissions in atmosphere and accidental leaks during transport, also them numerically measurable. The assessment of the environmental costs of transportation could suggest changes in the organization of the work, for instance in frequency and way of employee travelling and meeting, and for commuter travelling, with reference also to the impacts on the surrounding area and on the local traffic. Videoconferences could replace some travels, a company collective bus could be added to reduce the use of private vehicles, a larger flexibility in the entering and exiting hours would have positive effects on the local traffic in general.

Other critical environmental factors are those ones concerning waste, in their solid, liquid, and gaseous forms. The environmental cost connected with waste refers not only to the costs of collection, storage, transportation, treatment, and disposal, but also to the manners selected for these operations and to the possible alternatives, and involves some further assessments and splitting. For instance, it would be important not only to know the total amount of waste produced, but also the ratio between hazardous and non hazardous waste, because ways and costs of their management vary. Also the alternatives connected with the types of the selected treatment and disposal (for instance: composting, landfill, or incineration) and the quantities of reused or recycled waste inside the production cycle, or anyway in the company itself, influence costs.

About waste-water, further assessments should be added to the measurement of the concentration of the single pollutants referred to what is written in laws and regulations, and to the costs for collection, conveying, treatment and disposal. For instance, it would be useful to calculate not only the total amount of waste-water produced, but also the volumes sent to reuse and recycling inside the production cycle or inside the company itself. Moreover, the quality of the discharged water should be compared with the quality of the recipient water body, according also to the value that the local population gives to that resource, with reference to the kind of the final destination (for instance sewer or surface water). Also the type of the selected treatment influences costs.

About emissions in atmosphere, in addition to the measurement of the concentration of the single pollutants referred to what is written in laws and regulations, it should be assessed the total amount discharged in atmosphere and the relative amount connected with the ozone depleting gases, with the greenhouse gases, considered as the main causes of the climate change (particularly, the fluorinated gases are classified as high global warming potential gases according to the U.S. Environmental Protection Agency, EPA), and the relative quantities of NOx, SOx and small particles.

The environmental costs of materials and raw materials, contributing to the composition of the product, refer not only to the total quantity used but also to the relevant relative quantity of non renewable materials (minerals, carbon, oil and its by-products, natural gas, and metals), of by-product raw materials, and to the relevant relative quantity of materials reused or recycled in the production cycle or in the firm. In addition, the relative quantity of process or consumption materials, which do not enter in the composition of the product but are just needed for its production (for instance lubricants and cooling water), should be taken into account.

The raw material water can enter directly in the composition of the product or can be process water (for example cooling water). The environmental costs connected with the water supply concern the type of the supplying source, that can be an aqueduct, groundwater, or surface water. The assessment of the total quantity of water used in the production cycle should come alongside with the calculation of the relative quantity of water reused and recycled in the cycle itself, or anyway inside the company itself. In the assessment it would be important to consider also the quality of the supplying water, that is, for instance, if it is drinkable or if it is derived from a wastewater treatment plant. Potable water should be used only for food purposes.

Possible impacts on the water supply source (for instance a potential drawdown in the groundwater level) and in the area (for instance a reduction in the quantity of water intended for agricultural use) should be verified.

It is important also to assess what happens during the use of the product, for instance the presence of unpleasant sounds, the emissions in atmosphere, the consumption of materials and energy while working, the kind of the selected packaging (if it can be reused, recycled or divided into parts also reusable or recyclable) and so the cost of packaging disposal. The product itself presents some environmental costs too referred to its end of life destination, it could for instance be reused, recycled or divided into parts also reusable or recyclable, varying in this way its cost of final disposal.

Costs and numbers of the business impact on the surrounding natural non man-made environment can be both negative (loss, deterioration) and positive (revenue, improvement).

The negative aspect could concern the location of the company near to protected natural areas, the kind of the surrounding protected natural areas, the impacts of the organizational activities, products, and services on biodiversity, assessed considering the entire supply chain, the use of quarries, the potential pollutions, the potential effects on species of animals and plants protected and endangered for extinction, and the potential changes in the natural environment.

The positive aspect could concern the development of organizational strategies and policies directed to the adoption or to the restoration of a protected natural area, to the reduction of the impacts in order to improve its biodiversity, to the protection of species of animals and plants protected or endangered, to the collaboration with environmentalist associations for specific projects, and to the reduction of pollution.

The numeric assessment of the environmental sustainability in terms of energy refers to the measurement of the power consumption. Total consumptions should be divided on the base of the kind of the renewable and non renewable sources. Particularly, the use of non renewable energy involves a risk for the company of potential variability in the prices and in the supplies, that should be taken into account. The consumption of fossil fuels is also source of emissions of gases responsible for the greenhouse effect, causing the climate change. Moreover, energy directly used in the production cycle and energy used for services in the company should be considered separately. Also the energy technology has its own weigh in the assessment of the environmental sustainability, so as the interventions undertaken for energy saving, the technological improvements adopted, the interventions on the production cycle, the changes in the behavior of the firm personnel, the initiatives to improve energy efficiency, and the use of renewable energy.

All the data collected in that way should represent the business reality in a transparent manner, without omissions or manipulations, highlighting the positive results without leaving out the negative ones. It would be also necessary to explain in detail and accurately the basic hypothesis, the adopted methods for evaluation, calculation and measurement, which need to be all repeatable with similar results.

Data can be used to compare the results with those ones of the same company in different periods of time, to evaluate ongoing improvements or deteriorations, or for the same period of time with the results of different firms, also competitors.

The numerical assessment of the environmental sustainability of a company does not measure only its commitment for the protection of the environment, but describes a detailed picture to be used in the business decisions to optimize the production cycle, the organization and the management, the market goals, and the economic results in the long term.

What has been described for the business environmental sustainability can be similarly developed also in the social sector and in the economic sector.

Handbook for sustainable projects – Global sustainability and project management

3. Types of sustainable projects

3.1 Sustainable projects

Sustainable projects can be grouped in the following types, to which different levels of sustainability correspond:

1. Projects that are sustainable because of their own nature;
2. Projects that create sustainable products, results, or services;
3. Projects that are managed sustainably.

A project that is sustainable by nature is, for instance, a project for the remediation of a contaminated site. At a first analysis, this kind of project could seem sustainable only from the environmental point of view, actually it includes inferences from the social sustainability, such as improvement and restitution to the community of an area otherwise unusable, and from the field of the economic sustainability, with the subsequent reuse of the abandoned area and with the economic activities connected with the remediation itself.

A project of post disaster reconstruction is another example of project sustainable by nature. Its bond with the social sustainability is manifest and, if correctly managed, it could positively influence also the economic and environmental sustainability.

Among the projects for the sustainable development (Ministero dell'Ambiente, 2006) in Italy there are, for instance, those ones that include issues such as:

- Climate change;
- Trans-frontier pollution;
- Elimination of hazardous chemical substances and sustainable agriculture;
- Stratospheric ozone;
- Sustainable management of resources and sustainable development;
- Development of power sources and technologies with low emissions;
- Environmental education and information;
- Sustainable transportation.

Some international projects refer to the Alpine region in the following sectors:

- Management of natural resources;
- Climate change;
- Sustainable transportation;
- Renewable energy sources.

Some projects refer to the Mediterranean in sectors such as:

- Renewable energy sources;
- Sustainable tourism;
- Desertification;
- Sustainable transportation.

Examples of projects for the sustainable development in Middle-East Europe, Balkans, Central Asia:

- Environmental education and information;
- Renewable energy sources.

Examples in the wet zones of Mesopotamia, Iraq:
- Preservation of ecosystems;
- Desertification.

Examples in China:
- Monitoring and management of the environment;
- Environmental education and information;
- Protection and preservation of natural resources;
- Management of resources;
- Renewable energy sources;
- Sustainable urban planning;
- Sustainable transportation;
- Sustainable agriculture;
- Protection of biodiversity.

Examples in Central America:
- Climate change;
- Renewable energy sources.

Examples in Latin America:
- Sustainable management of forest estate for carbon absorption;
- Recovery of biogas from landfills;
- Environmental education and information;
- Assessment of negative impacts on the environment due to illicit cultivations;
- Development of methodologies and techniques for the recovery of deteriorated ecosystems.

Examples in the United States of America:
- Impacts of the climate change on the human health;
- Cycle of the carbon;
- Climate models;

- Low-carbon technologies.

Examples in India:
- Development of technologies for renewable energy;
- Forecast and prevention of climate change effects;
- Cycle of the carbon;
- Innovative technologies in industry and in agriculture for the reduction of the use of hazardous chemical substances.

Examples in Thailand:
- Prevention and management of risks for the coastal zones caused by anomalous events such as tsunamis and earthquakes.

Examples in Russia:
- Reduction of emissions through the development of advanced technologies for the use of renewable sources and for promoting the power efficiency.

Examples in Africa:
- Treatment of wastewater and re-utilization in irrigation;
- Integrated management of water resources;
- Management of the artificial recharge of aquifers;
- Exploitation of salt water and desalination process.

The World Bank provides funds for projects in the following sectors:
- Environmental protection;
- Sustainable development;
- Reduction of emissions;
- Acquiring carbon credits.

Afterwards, projects that create sustainable products, results, or services will be tackled in detail and, above all, a sustainable, organized and structured project management methodology will be provided.

In projects, which create sustainable products, results, or services, sustainability can be understood as a quality requirement and should be included in the technical specifications of the provided product, result, or service.

3.2 Globally sustainable projects

From the point of view of the global sustainability, projects, in a simplified and schematic manner, can be divided into:
- Traditional projects;
- Normally sustainable projects;
- Totally sustainable projects.

In the following schemes the adjective sustainable implies creating positive and lasting effects on triumvirate factors of environment, society, and economy.

From Figure 3.1, it is evident that the production cycle is different from the product, result, and service life cycle.

In the scheme of the traditional project (Figure 3.1), with raw materials it is meant enterprise organization, organization of the work, materials, energy, human resources, economic resources, and equipment.

Products can be products, services, or results.

In a traditional project the production cycle is linear, non continuous, and the roles of the organization, of the project manager, and of the final user are distinct and separate.

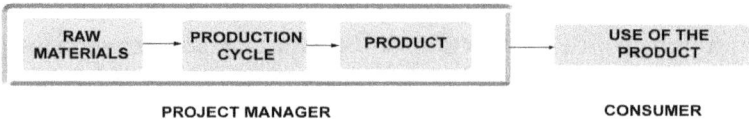

Figure 3.1 – Traditional project

In the scheme of the normally sustainable project (Figure 3.2), with protected and reusable raw materials available to the project manager it is meant:
- Ethical and enlightened enterprise organization;
- Positive and structured organization of the work;
- Reusable materials;
- Renewable energy;
- Other resources (economic resources, equipment) not to be impoverished.

In a normally sustainable project, towards which it should be aimed:
- The above written raw materials are all protected and reusable;
- The production cycle is respectful for environmental, social, and economic well-being;
- Products, services, or results are globally sustainable;
- Their use is globally sustainable;
- By-products are reusable outside or inside the production cycle itself;
- Waste is reduced to a minimum amount and is disposed of in a globally sustainable manner.

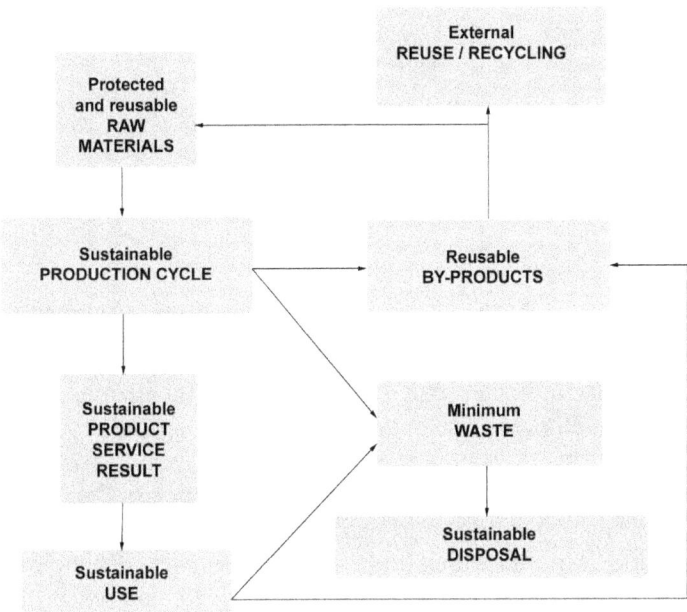

Figure 3.2 – Normally sustainable project

In the scheme of the totally sustainable project (Figure 3.3), not only the production cycle and the cycle of the use of the product, result, or service have to be continuous and non-linear to obtain sustainability, but sustainability itself must be simultaneously sought in its entirety and for each part of the whole.

Actually, it will not be possible to develop a project completely sustainable, or it will be possible only in a limited manner, because of enterprise, social, cultural, economic, and technological constraints.

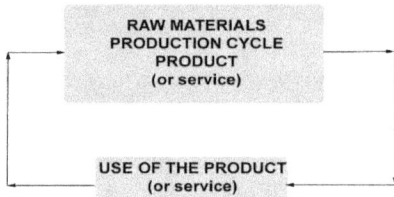

Figure 3.3 – Totally sustainable project

The ideal technologies could be not readily available, or they could be too much expensive or not yet much widespread.

Sustainability and innovation present a double bond. The research for sustainability often pushes to seek new affordable paths, that lead to innovations. At the same time, innovations, new technologies, materials, techniques, and instruments allow to add even more sustainability to projects, permitting a continuous improvement over time.

Project managers could have not much influence on the enterprise organization, on the choice of the raw materials and of the energy, and therefore on the real production cycle, likewise on the available equipment, on the economic resources, and on the selection of the suppliers.

Instead, directly and otherwise, project managers affect the work organization and the management of resources, particularly the human ones, and it is just in those sectors that this handbook can find wide application.

The sustainable management of a project implies an ethical and enlightened organization of the work, which would protect the human resources and would perform through daily personal actions. The sustainable project manager knows how to take care of all her or his environmental, social, and economic resources, and knows how to preserve their efficiency in the time.

3.3 By nature sustainable projects

3.3.1 Preliminary project of remediation of a contaminated site

The case study, described from the point of view of the designer and project manager, tackles the preliminary project of site remediation of an area of about 2.47 acres (10,000 square meters), included in a wider site of national interest in Italy.

It was a surface tuff quarry, partially filled with waste of different origin and characteristics, for a duration of nearly eight years, which polluted groundwater and soil. Both the quarrying and the filling activities were handled illegally and the site was under sequestration of the "Guardia di Finanza" (Italian police for economic and financial crimes).

Some dates:
1989 – Beginning of the quarrying activity;
1992 – Beginning of the filling of the quarry with waste;
2000 – End of the filling with waste;
2002 – Partial removal of waste;
2004 – Beginning of investigations and of the project.

The investigations, performed during the preliminary project of site remediation, allowed to estimate a total volume of waste of about 125,000 cubic yards (95,600 cubic meters), of which about the 80% formed by hazardous waste, and a volume of contaminated surrounding soil of about 119,300 cubic yards (91,200 cubic meters).

Considering the volumetric capacity of a truck of 105 cubic yards (80 cubic meters), about 1190 trucks were likely used to fill the quarry, which in 8 years of activity correspond to about 150 trucks per year.

The preliminary project of site remediation was developed according to the Italian law in force at the time (Ministerial Decree 471/99), which required:
  − Characterization plan;
  − Preliminary design;
  − Final (executive) design;
  − Site remediation works.

A generic WBS (Work Breakdown Structure) for a preliminary project of site remediation could include:
  − Geo-referenced topographic survey;
  − Geophysical explorations;
  − Soil drilling;
  − Installation of piezometers;
  − Hydrogeological investigations (flow tests, groundwater surveys);
  − Soil investigations;
  − Sampling;
  − Chemical, physical, biological, and radio-activity analyses;
  − Treatment tests;
  − Development of the preliminary design of the site remediation.

For the site of this case study, particularly, the activities of temporary dis-sequestration and issue of the permit for the access were added.

In order to develop the project schedule and to assign time, resources, and costs to the tasks identified in the WBS, it was necessary to know the internal resources of the company, that is the enterprise professional profiles and the relevant number of units and the availability of means and equipment. For what was not available inside, external suppliers would have been needed.

The organization had the structure of a weak matrix. The project manager had limited role and authority, was involved in several projects simultaneously, did not have control on the human resources, had limited economic resources, had difficulties in the relations with the other functional managers, above all because the latter worked in a different sector of the firm, that one of the special waste, considered as the strategic sector of the company. The technical director was in another office, at considerable distance from the operative office, did not have experience in the remediation of contaminated sites, and did not know the resources involved in the project. The working office and the working hours of the project manager and of the workers were different and differed also from the usual working hours of the suppliers in the workyard. No opportunities for career advancements or professional development were provided, both for the project manager and for the workers. The use of the project management procedures was appreciated, but not requested or adopted by the company. The firm adhered to the ISO 9000.

The workers assigned to the project were selected only by the functional managers, independently from the requests and from the needs of the project manager, and it was not possible to have their resumes and the description of their positions in the company. They assigned to the project: 1 full time surveyor, in the role of a coordinator, 14 part-time workers divided into two shifts a day, 2 part-time foremen, one for each shift, and their bus drivers. None of them had experience in the remediation of contaminated sites and they came from various sectors. The resumes were collected asking them directly to the workers, with the specific request to list also the "irregular" (without a contract) working experiences.

The surveyor had more than 20 years of experience in building yards and railway yards. The workers had an average of 37 years of past working activity, of which an average of about 46% (17 years) of various unemployment benefits, and therefore a period of effective work of about 54% (20 years), with an average of 36% (13 years) of irregular work, and so about only 7 years of regular work.

They were a very heterogeneous group. They had previously worked as workers in building yards, in the industry of metals, in the drilling sector, in the structural steel industry, they were textile workers, mechanical workers, marble-cutters, fur-cutters, electric welders and cooks.

Only with the help of an organized and structured methodology it was possible to obtain good results from the project. Practically, instruments, techniques, and procedures belonging to the project management were applied.

The first issue to solve concerned the members of the project team, which is to say workers and project manager, who did not know each others. The workers had a bad impression of their colleagues from the office, project manager included, and were suspicious and mistrustful. The adopted solution was to go every day on the site and to spend most of the time working with the workers in the field, with any weather condition, making possible a co-location in such way.

The second issue concerned the awareness that the workers did not have experience or training in the sector of the remediation of contaminated sites. The solution was to train directly the team about the job to be performed, to inform the workers about the main aspects of the job (technical specifications, laws and regulations, schedule, risks, agreements with suppliers, technological issues, relations with third parties, and use of the equipment), and to provide documents, memos, guidelines and forms.

The third issue concerned the project manager, who could not use wage increments or career advancements to reward the workers. The relations between the operative office and the workers were very difficult because of the enterprise policies. The project manager had no influence on the opinions, on the decisions, and on the enterprise policies. The project manager had no authority for the resolution of the problems or of the conflicts of the workers with the company. The workers had scarce means and equipment available. The solution was to admit to have the same problems of the workers and no authority or influence on the company, and to ask their collaboration and their help to solve together the common problems.

The fourth issue concerned the workers, who had rigid and unusual working hours, different from those ones of the project manager, and different from the working hours of the suppliers. The adopted solution was to delegate to the coordinator-surveyor the authority to make decisions during the absence of the project manager, according to the provided instructions and training.

The fifth issue concerned the workers, who could refuse to execute their job without incurring in meaningful consequences. The solution was to solve the previous problems.

The results were more than satisfactory. During the work on the contaminated site the workers were able to perform activities of cutting bushes, tracing the points contained in the drawings and in the maps, excavating by hand the test trenches, measuring the groundwater levels, making tests on the piezometers, both absorbing and pumping, decontaminating the drilling equipment, collecting the wastewater, and providing assistance during the sampling activities and to the suppliers. The coordinator-surveyor contributed also in managing the on site relations with the neighbours, with the inspectors for certification and control, with the local authorities, and with the police authorities. The whole project team contributed to the success of the project, working with commitment and in a professional manner.

The preliminary design of the site remediation ended in November 2005 and was approved by the Italian Ministry of the Environment in April 2006.

3.3.2 Projects for post disaster reconstruction

Disasters, both natural and man-made, are characterized by strong uncertainty, complexity, and emotional charge. They can be: earthquake, bradyseism, tsunami, inundation, epidemic, flooding, fire, landslide, volcanic eruption, and so on.

When a calamity happens, the phases of response and relief, generally assigned to the Red Cross, to the fire-brigade, to the army, to the humanitarian organizations, and the like, immediately follow. Very often the local resources, both human and material, have been themselves affected by the disaster, and help and interventions coming from outside the disaster area are needed. The calamity could also be of such dimensions, because of its extension or for its characteristics, to require the intervention of foreign countries.

During those early stages the reference is the Sphere Handbook, published in 2000 by the United Nations, available in several languages, and coming from a collaboration among International Red Cross, International Red Crescent, and various international non governmental organizations (NGOs). In the initial phases, of disaster response and relief, it is necessary to ensure that the living conditions of the victims of the calamity satisfy at least the Minimum Standards contained in the Sphere Handbook. A preliminary assessment of the disaster area is made to understand the impacts of the calamity on the health and on the livelihood of the affected population. The Sphere

Handbook suggests to make a general assessment for each technical sector inherent to:
1. Water and sanitation;
2. Nutrition;
3. Food;
4. Shelter;
5. Health.

Generally, the post disaster reconstruction starts from four to six months after the calamity and only when appropriate funds are available.

Projects for post disaster reconstruction, because of the characteristics of the calamities of uncertainty, complexity, and emotional charge listed above, need to be managed in a structured and organized manner to provide effective benefits. Their results, according to the project management definition of project, can be both products, for instance the construction of schools, dwellings, roads, hydraulic or sewer infrastructures, and services, for example health or psychological assistance to the victims of the disaster.

In these projects it is needed to follow the below list of priority:
1. Time;
2. Cost;
3. Stakeholders.

Stakeholders are all the parties, directly or indirectly, interested in the project and affected by the project, first of all the victims of the disaster, beneficiaries of the project. Their active participation is required in every stage of the project and is essential for the attainment of quality results.

Projects for post disaster reconstruction should create and build a bond with the area and with the population affected by the calamity. In fact, they are seen also as an instrument of in site aggregation and collaboration, to allow the victims to overcome together and united the difficult period.

It is essential to involve the disaster victims in the project, starting from the planning phase, not only to consider their needs, but also to collect more and detailed information on the local reality. In that stage, it could be needed the help of experts in the sector and of available documents on similar projects already developed, and about similar calamities already happened, to refer to certain and verifiable data and not to anecdotical information.

In a project for post disaster reconstruction a sustainability plan, which is a long term operational plan, should be developed (PMIEF®, 2005). While projects are by their nature temporary and have their own well defined beginning and end, their products, services, or results last over the time. When a project ends, it should be ensured that the disaster victims are able to continue to support themselves, making the impact of demobilization minimum, and giving continuity to what was provided at the completion of the project. If, for instance, health assistance has been provided following an epidemic, it could be thought to a vaccination plan in the long term view. Instead, if a building has been built, it could be planned for easy and inexpensive future activities of operation and maintenance.

Also the procurement planning should be done from the point of view of the sustainability of the area affected by the calamity. That means, for instance, to create work opportunities in the disaster area, to choose local suppliers or anyway suppliers appreciated by the local community, and to use material resources of the site.

A positive example of project (PM Network®, 2009) for post disaster reconstruction, and together of sustainable building, is represented by the construction of single dwelling units for public housing in New Orleans, United States, after the hurricane Katrina, which in 2005 affected the American Gulf Coast region. The new prefabricated houses are located in one of the poorest districts of the city, and mainly affected by the disaster. If necessary, they have the opportunity to float, rising up to twelve feet and remaining anchored. They have cisterns for the collection of rainwater, solar panels, equipment for water and power saving, high heat insulation, and the like. They have a certain water and power autonomy in case of necessity. They were built by the foundation "Make it right", funded by the movie star Brad Pitt, and have been designed with the involvement of United States university students and professors.

3.3.3 Projects according to the "Equator Principles"

The "Equator Principles" of the IFC (International Finance Corporation) of the World Bank are financial instruments in the sector of the renewable energy and tackle above all projects provided in developing countries, where the energy demand is more and more growing and where the power grid does not cover the entire territory. For instance, nowadays, in India about four hundred people live without electric power (PM Network®, 2011).

They are a set of volunteer norms to determine, evaluate, and manage social and environmental risks in project financing (Deshpande, 2011 and IFC, 2006). They have been adopted on June 4th 2003 following an agreement among ten banks, reached to protect investments from social and environmental risks.

They apply to all the industrial sectors and only for new projects, that include a no less than 10 millions US dollars investment. Till now, about seventy major banks and financial groups in the world adhere to these volunteer norms, representing about the 85% of the global market of project financing. While projects are executed above all in developing countries, banks have their head offices above all in already developed countries.

The management of social and environmental risks, from the very first stages of the project, limits the possibility that the project itself could be delayed, suspended, or cancelled. With this financial instrument, investors, to protect their own investment, want to be sure that projects would be executed in a socially responsible manner and according to the best and most updated practices of environmental management. Project financing does not allow investors, or allows them only to a limited degree, to recoup from the estate of the providers of the project, and therefore they have to protect themselves in advance from any danger that can threat the income connected with the accomplishment of the project itself. The investment pays back mainly with the cash flow of the project. For those reasons both the correct management of the project and the exact assessment of the social and environmental risks are important. It is the interest of everyone that the project would succeed.

Developing countries, growing continuously, need more and more energy. Where fossil fuels lack and the power grids are not much developed, and where the cost of the electric power itself is high, large projects are executed in the field of renewable energy, exploiting sun, wind, tides and geothermal energy to avoid importing gas, oil, and carbon. Moreover, where the power grid does not arrive, the distributed generation of power from renewable sources is needed. For instance, in India the International Finance Corporation has invested about a billion dollars for developing projects in the sector of renewable energy, and many other investments concern countries such as Morocco, China, Thailand, Philippines, Nigeria, Ethiopia, Kenya, and Rwanda. The most advanced countries in the use of renewable energy, and with the widest experience, are just those developing ones.

The Equator Principles of the World Bank divide projects into three categories, according as the potential social and environmental risk connected with them is high, medium, or low:

    A. Projects with averse, significant, numerous, irreversible, or without precedent potential social and environmental impacts;

    B. Projects with limited, in a scarce amount, generally site specific, mostly reversible, and quickly resolvable potential social and environmental impacts;

    C. Projects with minimum, or absent potential social and environmental impacts.

For projects with high and medium risk, a series of assessments, developed in such a manner and detail to satisfy the needs of the investors according to the peculiarity of the project itself, and the development of a risk management plan are needed.

The assessments can include safety at work, health and security of the population, evaluation of many alternatives, local and international laws and regulations in force, protection of human rights and cultural identities, protection of biodiversity, of protected species and areas, sustainable use and management of renewable natural resources, social and economic impacts, protection of ethnic minorities, correct use of dangerous substances, efficient use of energy resources, correct management of waste, prevention and control of any form of pollution.

Projects must comply with guidelines about environment, health and safety for the specific industrial sector.

For projects of categories A and B for developing or with non high income countries, so as grouped by the World Bank in specific lists, the development of an action plan and the determination of a social and environmental management system, including also mitigating and monitoring measures, are required. The action plan, according to the complexity of the project, could be formed by a single report, or by a set of many documents. The management system includes for instance the social and environmental assessment, management program, training, and involvement of the community for which the project is developed.

For high income countries the action plans are developed according to the relevant local laws in force.

For projects of categories A and B, concerning developing or with non high income countries, the Equator Principles require a public consultation to inform the population about the characteristics of the project and about the studies developed on its potential social and environmental impacts, and to collect, manage and solve potential disagreements and concerns raised in the community, for which the project is developed. This stage has to be managed through a specific structured plan that considers the needs of the population, the use of the prevailing local language, and the cultural diversities. This consultation stage, above all for the most delicate projects of the category A, should happen as soon as possible, and anyway much earlier than the project starts, and have to continue also during the execution of the project itself.

For the same above projects, a management mechanism should be included and made public to solve potential conflicts regarding social and environmental issues, that can worry or threaten single persons or groups of people belonging to the population affected by the project itself. The procedure should be simple, transparent, and rapid, and should be developed according to the local cultural traditions.

For all the projects belonging to category A, and if requested by the investors also for the projects of category B, an independent review of the documents from an external third party is required, to assess their compliance with the Equator Principles.

The financing depends also on the commitment to respect continuously a set of social and environmental laws and regulations. For all the duration of the financing, for projects of category A, and where requested by the investors also for category B, there will be an external expert in social and environmental issues, who will control continuously their compliance and will provide periodic reports about the proceedings.

All the projects financed by the International Finance Corporation are made public, excluding the confidential data, at least with yearly frequency, in order to provide practical examples about the adoption of the Equator Principles in executed projects.

4. Investments and projects

4.1 Difference between investment and project

The difference between investment and project comes out clearly from the above description of the Equator Principles. Projects are financed only if the investment satisfies particular characteristics. Investment analysis precedes project management.

For sustainable projects the investment analysis (Gareis et al., 2013) does not develop only according to the classical schemes:
- Economic assessment of the investment (the investment pays back mainly with the cash flow of the project);
- Cost-benefit assessment (study of the investment, developed in such a manner and detail to satisfy the needs of the investors);

but includes also:

- Social cost-benefit assessment (considering the impacts of the investment in the long term and on all the interested parties);
- Environmental impact analysis (considering the environmental impacts);

- Analysis of the context of the investment (considering the relations with other correlated investments).

Sustainable investments consider the point of view of all the parties interested in the project and affected by the project.

While the project is temporary by definition, the investment to be sustainable should be analyzed in the long term.

Some authors (Carboni et al., 2013) suggest to develop a sustainability management plan to be considered in the phase of the analysis and of the evaluation of the investment, above all for its impacts on the management of the resources and on the risks of the project. This denomination actually generates confusion, because the management follows the decisional stage and comes alongside with the development of an already approved project. In the phase of evaluation of an investment, it would be more correct to speak of a Sustainability Charter, like that one adopted by the North Vienna Hospital (Vienna Hospital Association, 2010) and only later integrated in the approved project (or better, projects and program) management plan.

A project starts only when the investment has been analyzed and approved, when the enterprise organization that will execute it, has been selected, and when the project manager has been officially appointed. The document that ratifies the beginning of a project is normally called project charter.

An example of social cost-benefit assessment is provided below.

4.1.1 Social use value of environmental goods

The social use value of environmental goods integrates all the three aspects of sustainability: economic (value), social (social use) and environmental (environmental goods).

To develop an economic analysis of an investment we need to estimate its economic value and the same applies also to investments in the environmental field. The economic analysis refers to the community, from where it links with the social aspects, and differs from the financial analysis, which on the contrary pertains to the single private operator, who undertakes it (Forte, 1977).

The social use value (or economic value) of an environmental good refers to the appreciation that the community has of it in terms of its collective usefulness and usability. The assessment of the social use value involves also some factors that are not easily quantifiable in terms of economic efficiency, which cannot be defined in terms of profitability, are not directly expressible in money, and, as such, fall under the group of the so-called "intangibles".

Before continuing the discussion, some brief and non exhaustive references of microeconomics are needed. The economic good (commodity) is a product or a service, which has characteristics of utility, usability and limited availability. It can be durable or nondurable, present or future.

The exchange value or market value is the most likely value, expressed in money, of a commodity traded in a market and is a historical datum related to the particular market. The cost value is the sum of the market values of all the inputs needed for the production of the good. The production is the transformation of natural or material goods into economic assets of greater utility. The complementary value is the cost value plus the depreciation. The transformation value is the market value after the transformation, minus the cost of the works required for the transformation. The value of subrogation is the market value of another commodity with the same utility. For full details on the above definitions and about the estimation methodology, see the reference at the end of the book (Orefice, 1984).

The value of the environmental goods arises when natural goods such as water, air, soil, land and the environment itself generally speaking turn into economic goods. They are no more unlimited resources and their availability becomes limited in quality or in quantity. From the open economics of the cowboy we move to the closed economics of the spaceship earth (Boulding, 1966).

In the old days, in conditions of deterioration of the natural environment or of the social structure, there was always the possibility of conquering new spaces more livable on the planet. Now, on our planet there are no more new territories to discover and to occupy or frontiers to pass. The earth is no more a limitless expanse, but a finite sphere. Today, also the economics must look at the planet no more as an open system, but as a closed system, where everything takes place inside it; inputs (for instance energy and raw materials) come from inside and also the outputs (for example waste and pollutants) remain inside. Consumerist mechanisms of production, which create unmanageable waste and are based on planned obsolescence, fatuous and unfounded needs, products of poor quality and misleading advertising, cannot guarantee a sustainable future, the survival and the preservation of the human species. A sustainable economics should not only satisfy the urgent needs of the market, but should always consider the needs of the future generations and ensure the transition from an anthropocentric to a biocentric vision.

The clean, renewable energy is today the new frontier of the global economic growth. Also the gross domestic product GDP, a measure of the economic success of a country, in a modern economy should be divided into two parts, one coming from renewable energy and resources and the other one resulting, on the contrary, from non-renewable energy and resources.

The innovation in the environmental field projects towards the research of production processes and technologies of high energy efficiency, which reduce the negative externalities associated with the production of waste and with the pollution (Nyangon, 2011). Continuous production cycles where the end of life of an old product coincides with the beginning of the life of the new product, and no more the linear ones should become increasingly the norm and not an exceptional example. From the economic point of view, a new appreciation of the profits coming from the production of durable goods would be needed, shifting the attention from the revenues of the mere sales of the consumer goods to the earnings connected instead with the services for the operation and the maintenance of the products themselves. Global sustainability has also become an essential requirement for financing many projects in the environmental field (see for instance the Equator Principles).

In light of what written above, some considerations about possible methods of assessment of the social use value of the environmental goods follow (Forte, 1977).

The exchange value or market value (Vm) of a good is the most probable value, expressed in money, that a single operator would be willing to pay in a given market for that good. The social use value (Vsu) differs from it since it does not concern a single operator, but the appreciation of the entire community (economic value) for that same good. This (positive or negative) difference is the social surplus value of the good ($\Delta V = Vsu - Vm$).

In the field of the "intangibles" we have to consider the direct benefits to the community caused by the environmental goods and that cannot be expressed in terms of economic efficiency. Among them we can consider the benefits to the reputation of the place, to the civil and cultural development, to the creation and the sharing of knowledge, to the transmission of historical memory to the future generations, to the health, meant as a state of complete physical, mental and social well-being, as defined by the World Health Organization. In this case the social use value could be determined with the so-called "willingness to pay", interviewing a selected and significant sample of potential users of the environmental good. The results of a similar survey would be anyway susceptible to subjective evaluations.

A possible assessment of the social surplus value ΔV of the environmental good could come from the *discounting of the annual expenditure S*, discounted at the rate r of the social return, for its preservation, maintenance, vigilance, protection and promotion.

$$\Delta V = S / r \quad ;$$
$$Vsu = Vm + S / r \quad .$$

In this calculation the rate of return on a public and social investment (rate of the cost of social opportunity) will be necessarily less than the financial rate of return on a private investment.

Another method of estimation is the *"shadow price"*, Ps, or money of account. The shadow price is the price that reflects the social value of a good or of a service of which there is no market.
It can be calculated using one of the following ways:
- Applying the actual prices that occurred in other markets to benefit of the same goods and services;
- Calculating the *external effects* with reference to the market prices;
- Estimating the shadow price with a political decision, consistent with the defined objective.

In the first case, not always the Italian environmental goods are comparable to those of other countries; we can consider, for instance, the archaeological and natural underwater parks of Baia and of the Gaiola, the caves of Naples underground or the Fontanelle Cemetery.

In the second case, *external effects*, only the increases of the income from tourism are easily assessable. An estimate of the social surplus value of the environmental goods, in terms of economic efficiency, could be obtained discounting at a reasonable rate of social return the net increase in the income, national or local, coming from the portion of the total annual tourist flow induced by them. The environmental tourism and that one linked to the natural locations of the movie sets are now very popular.

Conversely, it is hard to estimate the increase in social well-being achieved with the use of the natural good. An assessment of the increase in the pursued level of education could take into account the value of the courses offered on the environmental goods and in the environmental goods themselves, books, publications and documentaries, seminars and conferences, theater performances and concerts organized in forests, coastal dunes, or in underground caves. Many are the courses and the cultural activities for instance on subjects such as speleology, rock climbing, trekking, mountaineering, scuba diving, botany, evolution of species, biology, geology, biodiversity, astronomy, photography and the like.

They could also be evaluated and taken into account the avoided environmental damages, and thus the costs saved, with the preservation and the protection of the environmental resources, such as the economic damage to a part of a coast no more swimming because of the pollution caused by the discharge of untreated waste water.

In the third case, political decision, the social use value would come directly from the importance that the community, through the political class that democratically represents it, assigns to the estate of the environmental goods. If, for instance, among the political objectives there is the improvement of the quality of life, then the fruition and the protection of the environmental goods will be appropriate means for pursuing it and the environmental good will have for the community a use value, or economic value, greater than its exchange or market value.

The social use value of the environmental goods is closely related to the global sustainability.

5. Global sustainability and project management

5.1 How to manage a project sustainably

The best way to manage a project sustainably is:

- Following a standard;
- Including the sustainability objectives from initiating.

A referring standard could be the 2013 ISO 21500 *"Guidance on Project Management"*.

Five process groups and ten knowledge areas can be distinguished, as described in Table 5.1.

According to the specificities of the project and to its characteristics, the necessary processes are selected among those ones available in the referring standard.

Table 5.1 – Process groups and knowledge areas

5 process groups	10 knowledge areas
1. Initiating 2. Planning 3. Executing 4. Monitoring and controlling 5. Closing	1. Integration 2. Scope 3. Time 4. Cost 5. Quality 6. Human resource 7. Communications 8. Risk 9. Procurement 10. Stakeholder

For the project success, it is necessary that the sustainability objectives would be already included in the documents that initiate the project itself (for instance the project charter). This means that the sustainability objectives have already been considered in the investment analysis and that the project manager and her or his project team have been selected, according to their specific expertise in the sector, and would know and agree on its basic values, agreed also by the organization itself, and included in the business strategies. The fundamental values are written and described for instance in GRI (2011), ISO 26000 and UN Global Compact (2011).

The principles of the sustainable development (Gareis et al., 2013), which provide a holistic vision to the project, are:
- − Environmental, social and economic sustainability;
- − Short, medium, and long term perspective;
- − Local, regional, and global perspective;
- − Agreed basic values.

Among the fundamental values, together with the respect for legality and for an ethic code, there is for instance the awareness of the need of a wider participation of stakeholders, and of the need to give a wider decision-making authority to more subjects (empowering) that would imply a certain degree of autonomy and of responsibility, avoiding organizational charts with power centralization.

Among the principles of sustainability (Silvius et al., 2012) there is also that one of consuming the income and not the capital. It means that resources should not be consumed faster than they could be regenerated. The concept applies also to the human resources, important organizational asset, that must be preserved over time.

The ethics of the sustainable project manager (Silvius et al., 2012) is based on an ethical, or conduct, code and includes doing choices or making decisions that would find a right balance among the social, economic and environmental aspects, respecting the sustainability principles and the organizational strategies.

The project manager, before an ethic dilemma, can select two kinds of approach:
- Utilitarian ethics;
- Deontological ethics.

Utilitarian ethics looks at the future consequences and at the greatest future utility of all the actions, which is possible to undertake in that circumstance for all the stakeholders of the project.

Deontological ethics refers to moral laws agreeable by anyone, which should be universal natural moral laws, not influenced by our senses and by our traditions (Silvius et al., 2012, citing Immanuel Kant, 1797).

Referring to global sustainability, any available decisional ethic instrument would have been chosen to be used (for instance PMI®,

2012), the suggestion is to consider some absolute values, independent from the place, that could be everywhere, or from the time, that could be also future, in the analysis of possible scenarios and alternatives.

5.2 Project charter

As highlighted in the previous section, the best way to manage a project sustainably is that the sustainability objectives would be already considered in the investment analysis and included in the initiating stage.

The project can be:
- Sustainable by its own nature;
- Providing sustainable products, results or services;
- Managed sustainably.

In the projects providing sustainable products, results or services, sustainability can be understood as a quality requirement and should be included in the technical specifications of the provided product, result or service.

In the prosecution it will be referred to a project that provides sustainable products, results or services, and that is managed sustainably.

In the development of the project charter the description of the project should include, in a summarized but exhaustive manner, the general purpose of the project, the description of the sustainable products, results or services, which are intended to be provided both at the end of the project and at the intermediate milestones, the WBS, the sustainable use of resources, the schedule of the project and the scope of the project, that is what is included in and what is excluded from the project.

In the purpose the reasons, which drove to the approval of the economic investment, and the parties receiving benefits from the project should be indicated. Data on the investment analysis should be provided.

Project goals should include environmental, social and economic sustainability objectives, in the short, medium and long term, on a local, regional and global extent, and they should be specific, measurable, agreed, realistic and allocated in the time.

Project stakeholders, their needs, and their requests should be identified and described. Stakeholders can vary proceeding the project over time. Stakeholders are also future generations. Stakeholder needs and requests should be assessed in the environmental, social and economic areas, in the short, medium and long term, on a local, regional and global extent.

In assumptions and constraints the adopted sustainability criteria should be indicated, for instance constraints could be imposed on the selection of the suppliers, satisfying particular sustainability requirements, such as the signature of the UN Global Compact. In the assumptions the agreement on the basic values of sustainability among all the parties involved could be included.

In the assessment of risks and opportunities, both project risks and opportunities and stakeholder risks and opportunities in the environmental, social and economic areas, in the short, medium and long term, on a local, regional and global extent should be considered.

Project supplies should be described in their characteristics of time (milestones, deliverable time), cost and quality. Particularly, the sustainability requirements of quality should be added to the technical requirements. The description of products, results or services should be sufficiently clear and understandable, so that the project team could provide them correctly and accordingly to the agreements for their delivery. Moreover, what is included in the supplies and what is excluded from them should be indicated.

The project schedule should consider both the final deliverable and the intermediate deliverables or milestones. A project, which takes into account also the future generations, should include some considerations in the long term, that go beyond the deliverables themselves and the time of the project itself, and that involve also the use of results, products or services provided with the project and their final destination at the end of life.

The resources used in the project are economic, material (raw materials, energy, equipment, supplies, services), and human (employees, consultants, contractors). They should be indicated in quantity, quality, and allocation in the time. The management of all the resources should respect the sustainability requirements. The choices about the use of the resources will determine the costs of the project and their distribution in the time.

The project charter should contain also the criteria for evaluating the success of the project, that is if the project objectives have been met in a satisfying manner. For this reason all the project objectives should be measurable over time. Also the benefits of the project should be determined and assessed in terms of global sustainability.

The project objectives can refer to the different knowledge areas, as described for instance in Table 5.2, where risk + indicates an opportunity and risk − indicates a negative event for the project itself. Each objective should have its own allocation in the time and in the space, and should provide a range of values considered as acceptable.

Further specific directions about the sustainable management of the project can be added to the project charter according to the particular examined context.

Table 5.2 – Project objectives by knowledge areas

Knowledge area	Economic objectives	Values and measurements	Environmental objectives	Values and measurements	Social objectives	Values and measurements
Scope						
Time						
Cost						
Quality						
Human resource						
Communications						
Risk +/-						
Procurement						
Stakeholder						

5.3 Project management plan

The project management plan includes plans such as the risk management plan, the communication management plan, the stakeholder management plan, the procurement management plan, the scope management plan, the change management plan, etcetera.

Some authors (Maltzman and Shirley, 2011) suggest to include a separate plan for the environmental management (environmental management plan). Actually, as Gareis et al. (2013) highlight, if the objectives of global, not only environmental, sustainability are added to the initiating documents of the project, these parameters will flow automatically into all the documents that later will form the traditional project management plan.

Some of the documents, which will compose the project management plan and which are significant and innovative examples in their aspects relevant to sustainability, are the logistic plan and the sustainability plan.

5.3.1 Logistic plan

The logistic plan contains policies, procedures and guidelines to coordinate the material and human resources and to ensure that they are handled sustainably in order to meet the requirements of the project. It describes the activities of coordination and synchronization of transportation and of movements of people and goods under the best conditions of efficiency, considering their implications in the environmental, social and economic field, in the short, medium and long term, locally, regionally and globally.

It is closely connected with the human resources management plan, with the procurement management plan, and with the schedule of the project activities.

5.3.2 Sustainability plan

Similarly to what is done in the projects for post-disaster reconstruction, it would be appropriate to develop a sustainability plan, intended as a long term operational plan. While projects, because of their own nature, are temporary and have well-defined beginning and end, their products, services or results last over time. When the project will be completed, it should be ensured that all the users of the product, service, or result are able to manage it sustainably in its economic, social and environmental aspects.

It can be thought for instance about the supply of durable household appliances with low water and energy consumption, easy in their operation and maintenance, and easily recyclable or disposable at end of their life cycle. Another example could be the supply of a service, that creates, preserves and increases the job opportunities also at the end of the project. Another product could be a building designed and built to last over time, accessible and usable for people with disabilities, already prepared for possible changes or extensions, and that allows easy and inexpensive operation and maintenance.

This would mean to work from the planning and execution stage to deliver a product, service or result provided with an instruction "manual" for future sustainable use and maintenance, giving continuity to what has been delivered at the end of the project.

5.4 Time

Time is a concept of fundamental importance in a sustainable project.

For sustainable projects different time can be described, such as:
- − Time of the project;
- − Time of the product, service, or result of the project;
- − Time of resources;
- − Time of raw materials;
- − Time of energy;
- − Time of the organization;
- − Time of stakeholders.

Reserves of some raw materials renew in time corresponding to geological eras, for instance oil and its by-products, other resources instead renew in a short time, for instance water, through biogeochemical cycles.

In the time of stakeholders also future generations should be considered.

Since among the principles of sustainability (Silvius et al., 2012) there is also to consume the income and not the capital, all resources should not be consumed faster than they can be regenerated, even the human ones.

Time is then no more just a concept related to the mere development of the schedule of the activities, but it concerns the ethical management of human and material resources in its impacts in the environmental, social and economic field, in the short, medium and long term, on a local, regional and global extent, and what happens after that the project has been completed. The time beaten by the classic schedule of the project should consider time and duration of the activities identified in the WBS (generally indicated in weeks), the days of the intermediate deliverables, the days of the intermediate milestones (deliverables and milestones are identified through a date in the schedule, having no duration), the day of the final deliverable, and also the time of the sustainable management of resources.

A project that would take into account also the future generations should include considerations on the long term, which go beyond the deliverables themselves and the time of the project itself, and which involve also the use of results, products and services provided with the project and their last destination at the end of the life cycle.

While the project is in progress and at its end, the respect for the sustainability objectives identified for time in the initiating stage of the project should always be verified.

5.5 Cost

Costs are greatly influenced by the sustainability principles, for instance a rational and sustainable use of resources could provide savings in the cost of energy, of fuels and of materials, an optimization of time, and a better organization of work.

The reduction of waste from production and its correct management in treatment and disposal could lead for example to significant cost reductions. The use of renewable energy could protect against the risk of potential variability in prices and supplies.

Also for the cost the effects should be considered in the environmental, social and economic field, in the short, medium and long term, locally, regionally and globally.

A project has not only economic costs, but also social and environmental costs which, if not calculated or if neglected, will go later to burden on society or, in the long term, on the same company providing the project. Giving priority only to the economic aspects related to the profitability of the project, it is implicitly decided to transfer its environmental and social costs on the community, for example with a nonchalant waste management or without adopting the individual safety devices. Actually, it is just a transfer of costs, since all the issues and responsibilities that are not faced with the project, then glide over the whole community or rebound back on the company itself.

A sustainable approach, replacing a traditional approach, with its comprehensive analysis could lead to the evaluation of new possible, not yet explored, alternatives for each of the activities included in the WBS, suggesting potential decreasing in project costs.

The cost estimate should be developed in detail for each activity or milestone listed in the WBS and in the schedule, written also in accordance with the sustainability criteria. Among the items to be considered there are the cost of personnel, of materials, of energy, of travels, of waste, of communications, of suppliers, etcetera. The cost of the transportation of the human resources, of travels and of communications can certainly be contained and reduced with the information technologies today available and under continuous innovation. From the aggregation of the single items, then the overall cost will be estimated.

These general guidelines should be detailed depending on the particular project developed, on the specific sector of activity and, if necessary, with the advice of one or more experts.

While the project is in progress and at its end, the respect for the sustainability objectives identified for cost in the initiating stage of the project should always be verified.

5.6 Quality

Quality may concern:
- The sustainable management of the project;
- The sustainability of the result, product or service provided with the project;
- The sustainable production cycle of the product or result, or the sustainable manner of the service delivery.

As previously highlighted, in projects that create sustainable products, results or services, sustainability can be understood as a quality requirement and should be included in the technical specifications of the provided product, service, or result.

The deliverables of the project should be described in their characteristics of time (milestones, deliverable time) and quality. Particularly, sustainability quality requirements should be added to technical requirements.

A sustainable production cycle is a continuous and non-linear cycle. An example is that one proposed by the C2C® methodology of the so called Cradle to Cradle® design (Braungart and McDonough, 2002), created by Michael Braungart and William McDonough to design new processes, products and services, eliminating the concept of waste, using energy from renewable sources, mainly from the sun, and promoting the cultural and biological diversity. The proposed model aims to restore the continuous cycles of nutrients, both biological and technological, with positive effects in the long term on organizational profits, on the environment, and on the human health.

The quality objectives of the project should include environmental, social and economic sustainability objectives, in the short, medium and long term, on a local, regional and global extent, and should be specific, measurable, agreed, realistic and allocated in the time.

While the project is in progress and at its end, the respect for the sustainability objectives identified for quality in the initiating stage of the project should always be verified.

5.7 Human resources

The human resource management depends mainly on the organizational culture and structure and is based on the respect for some values known as corporate social responsibility, such as the respect for the rights of the workers, the well-being of the workplace, the sustainable management systems of human resources, the exclusion of child labor, forced or compulsory labor, the respect for health and safety in the workplace, the freedom of association and the right to collective bargaining, the opposition to discrimination, the sustainable working hours, the existence of disciplinary actions, the assurance of a fair remuneration (SAI, 2008). These basic values should also be agreed with suppliers, contractors and subcontractors.

The human resource management depends also on working environment, geographical location, communications, organizational internal and external policies, cultural issues, ethical and professional behavior, interpersonal skills. It can be analyzed in its economic, social, and environmental implications.

The economic aspect is strictly related to productivity, seen as the efficiency of the factors used in the production process, and to avoid global waste of money. It concerns for instance the correct management of: wages, virtual working teams, electronic communications, professional training and certification of personnel, bonuses, rewards and awards, security, location of the workplace (for instance proximity to public transportation), perquisites (for example the use of the firm's car), availability of business services (such as parking, restaurant, nursery, gym), organization of commuting workers, policies of personnel administration, use of information technologies for meetings, events, travels, job interviews, recruitment procedures, and improvement of interpersonal relationships.

The social aspects are connected with: equity, workers' health in the long term, quality of life, improvement of interpersonal relationships, quality and improvement of the working environment, well-being, education, moral values, personal skills, social inclusion, adaptability, sufficient income for employees to economically support themselves and their families, personal learning and development, family services, workplace safety, security, opportunity to develop and improve personal skills, ethical behavior, ensuring respect for all human rights and rights of the workers, transparency, honesty.

The environmental aspects relate to: use of clean and renewable energy, control and reduction of waste, appropriate collection, storage and disposal of the residual waste, reuse, recycling, energy and water saving, water reuse, quality of the working environment, wise use of natural resources, prevention of pollution, protection of air, water and soil, knowledge and improvement of the natural environment.

Some of the aspects mentioned above, such as the quality of the working environment, appear repeated many times in different contexts, reflecting the fact that the environmental, social and economic components of sustainability are strictly interlinked to each other.

As previously written for cost, also for the human resources, giving priority only to the economic aspects, it is implicitly decided to transfer their environmental and social impacts on the community. It is actually just a share or a transfer, because the relevant issues are not addressed and the responsibilities glide over the entire community or single individuals.

For example, if the continuous training of personnel is not ensured by the employer, the worker could be forced to obtain the professional certifications at her or his own expenses or, if the workplace is not healthy and safe, the health care costs could indirectly be paid by the whole society, or poor working conditions could have negative impacts on the social relations and fall on the whole community.

A good project human resource management could lead to conditions of global sustainability for the society as a whole.

Within a given organizational context the project manager could adopt some tools and techniques for a sustainable management of the human resources assigned to the project.

The project team can be managed sustainably (Silvius et al., 2012), developing its strengths and managing positive goals, positive emotions, and positive relationships.

If part of the staff is inexperienced, some processes will be needed to develop the project team. Emotional intelligence skills and specific training, also on the job, could be used. Some basic rules of conduct, taking into account the cultural differences, should be identified, planned and defined, and awards should be given at the attainment of the project objectives, maybe at particular milestones or deliverables. The dynamics of the project team are also influenced by personal and cultural differences in the work style. It could be necessary to clarify roles and responsibilities, also in relation to the defined sustainability goals, and to manage a potential scarcity of resources. If that is not enough to create harmony in the working team, tools and techniques for the resolution of possibly arisen conflicts should be used.

It should be highlighted that the responsibility for the achievement of the environmental, social and economic sustainability objectives of a project should be clearly, and in writing, assigned to specific personnel. In the documents that describe the human resource need, in compliance with the schedule of the activities and the worker flows, the type of required resource should be identified and, in the description of roles and responsibilities, it should be indicated if and what responsibilities exist in the area of sustainability. The overall responsibility of the project is of the project manager, but specific activities may require particular skills and responsibilities. If the required resource is not already available in the company, or cannot be trained inside the organization, it must necessarily be acquired outside and will involve the procurement management.

In the example of the preliminary project for the remediation of a contaminated site (section 3.3.1), the training on site of the workers, the clarity and the transparency, the attribution of decision-making authority, involving a certain degree of autonomy and responsibility, all techniques characteristic of a sustainable management of human resources, have contributed substantially to the success of the project itself.

While the project is in progress and at its end, the respect for the sustainability objectives identified for human resources in the initiating stage of the project should always be verified.

5.8 Communications

The project communications management relates to the management of e-mail, phone calls, meetings and the like with all the project stakeholders, to inform them about the state of the project itself and to resolve potential issues. All communications should be open and honest.

The information to be distributed may concern the project status, changes, and deadlines. In the communications management plan what information is needed, to whom it must be notified, in which form, in which detail and how often should be indicated. A schedule for the information distribution, depending also on the duration of the project, should be developed. Some parties will be interested in a general overview on the progress of the project, others will need instead indications about the technical details and about measures on the achieved objectives. Some stakeholders will prefer to be informed in person, for others for instance press releases will be needed. For other subjects the tele or video conference could be required.

Also for project communications the criteria of environmental, social and economic sustainability, in the short, medium and long term, on a local, regional and global extent should be followed.

Choosing appropriately among the available and continuously innovating technologies, it will be possible to limit and to reduce the costs of the communications and to optimize their time and quality.

As an example, Table 5.3 is shown.

Table 5.3 – Information distribution

Stakeholders	Which information	In which form	In which detail	With which frequency
Stakeholder 1	Attained results	In person	Measures	Weekly
Stakeholder 2	Technical issues	E-mail	Technical	Daily
Stakeholder 3	State of the project	Meeting	General report	Monthly
Stakeholder n	Sustainability objectives	Video-conference	Measures	Bi-weekly

While the project is in progress and at its end, the respect for the sustainability objectives identified for communications in the initiating stage of the project should always be verified.

5.9 Risks

The definition of risk from PMBOK® (PMI®, 2013) is: *"Uncertain event or condition that, if it occurs, has a positive or negative effect on one or more project objectives"*.

In the assessment of risks and opportunities both risks and opportunities for the project, and risks and opportunities for stakeholders in the environmental, social and economic field, in the short, medium and long term, on a local, regional and global extent (Gareis et al., 2013) should be considered.

Impact and probability of occurrence should be considered in the qualitative risk assessment and for the composition of the risk matrix.

An example of a criterion for the evaluation of the weighed impact of the risks on environmental, social and economic sustainability is given in Table 5.4. The impact can be measured as high (10), medium (6) or low (1), while the sum of the relevant weights is set equal to 10. Risk + indicates an opportunity and risk - indicates a negative event for the project (P) and for stakeholders (S).

Table 5.4 – Impact of the project risks

Risk	+/-	P/S	Environmental impact	Environmental impact weight	Social impact	Social impact weight	Economical impact	Economical impact weight	Weighed impact
Risk 1	-	P	8	4	4	4	5	2	58
Risk 2	+	P	2	5	5	2	6	3	38
Risk 3	-	S	8	1	6	1	4	8	46
Risk 4	-	P	1	2	2	6	2	2	18
Risk 5	-	S	7	4	3	5	5	1	48
Risk 6	+	P	9	7	6	1	8	2	85
Risk 7	+	P	3	4	7	3	6	3	51
Risk 8	-	S	6	3	2	2	6	5	52
Risk 9	-	P	4	1	9	4	3	5	55
Risk 10	-	P	2	5	4	1	1	4	18

The probability of occurrence is the probability that a risk will occur. Both, probability and impact, refer to specific events and never to the project as a whole.

For each item it is necessary to identify what are the actions in response to risks and opportunities, when, how and by whom they should be implemented.

The identification of the risks, the qualitative and quantitative analysis and the management of the risks depend on the organization culture, policies and strategies, on the natural, political, technological and social environment, on the financial and economic situation, on the conditions relating to human health, and on social security.

In the identification of the risks, in addition to the traditional tools and techniques as for instance the so-called "brainstorming", the systemic constellations can be used, because a project may be considered as a social system (Gareis et al., 2013). Brainstorming is a technique of group analysis, where the research of the solution to a given problem happens through the free exposition of ideas and proposals from all participants in a meeting (Zingarelli, 1988). The systemic analysis studies living, socio-economic and material systems, in mutual relation among them, considered as physical entities formed by interdependent elements represented through symbolic objects (in the cases studied they are represented with pieces of wood of different colors and shapes) placed on the work table, analyzed and interpreted in their interrelationships through group dynamics with the aid of an expert in this technique.

In the risk identification the widest participation of stakeholders is required.

The risks, identified in their order of priority, should be constantly monitored on a daily, weekly or monthly base, depending on their weight on the project or on the stakeholders. For the activities deemed as critical, or for which the associated risks are occurring, monitoring should be more frequent.

In the project documents, for each main risk or opportunity identified, they should be clearly indicated the owners of the risks, the agreed responses to the risks, the material and human resources to be used, the warning signs of the risks, the residual and secondary risks, and the contingency reserves for schedule and cost. It should also be developed the check list for low priority risks.

Among the documents for the management of risks, there are also the security plan and the contingency plan.

The security plan concerns the responsibilities of the company and of the project manager towards the workers and refers to the protection of well-being, health and safety in the workplace, and to the personal security. It could consist of a so-called HSE (Health, Safety and Environment) plan adopted especially in industries, it could be a plan of vaccinations for humanitarian interventions in areas with epidemic outbreaks in progress, or it may consist of an evacuation plan in an area affected by a calamity. It depends above all on the particular sector of activity of the company providing the project, and is usually a general plan of the firm and not a specific plan for the project.

The contingency plan is a specific plan for the project and refers to the response plan to the risks for the project or for the stakeholders, in case they occur or are expected to occur. It should take into account the entire management plan of the project and particularly the impact on cost, schedule, human resources and procurement. It should describe the selected actions to accept the risks or to mitigate them, reducing their impact or their probability, or both. It should identify the warning signs of the risks (risk triggers, initial causes or symptoms), because the occurrence of an initial cause, while not necessarily implying that the risk will occur, increases its probability of occurrence. It should describe what actions have to be undertaken (including how and by whom) when a risk or its initial cause occurs.

In Table 5.5, a schematic example of the management plan of a single risk is shown.

While the project is in progress and at its end, the respect for the sustainability objectives identified for risks in the initiating stage of the project should always be verified.

Table 5.5 – Management of a single risk

Management of risk 1	Action
Risk description	Describing in quality and in quantity the risk, its probability of occurrence and its impacts on the project or on the stakeholders in the environmental, social and economic areas, in the short, medium and long term, on a local, regional and global extent
Risk trigger	Indicating in quality and in quantity the first cause and its effects on the increase of the probabilities of occurrence of the risk
Contingency plan	Describing the response plan to the specific risk
Risk owner	Indicating the resource responsible for the management of the particular risk
Report on the risk managed	Describing in which form, in which detail and with which frequency the risk owner will provide a report about how the risk has been managed (report, meeting, telephone call, e-mail, frequency, measures done, technical details, collected data)

5.10 Procurement

Procurement should be planned with the development of clear and detailed technical specifications, which will help to select qualified suppliers and to sign contracts.

The contracts will be written according to the local culture and customs and will comply with in force laws and regulations applicable to the specific sector.

Also for the supplies the sustainability objectives in the environmental, social and economic area, in the short, medium and long term, on a local, regional and global extent should have been chosen.

If one of the selected objectives is to promote local sustainability in the long term, it would be needed for instance to create job opportunities on site, to choose local suppliers and to access to local resources.

Among the principles of sustainability (Silvius et al., 2012) there is also to consume the income and not the capital. It means that, in general, resources should not be consumed faster than they can be regenerated.

In the assumptions and constraints of the project, indicating the sustainability criteria adopted, for instance, some restrictions could have been required for the selection of suppliers, contractors and subcontractors so that they meet specific sustainability requirements, such as adopting the C2C® model of a continuous cycle for the entire supply chain (Braungart and McDonough, 2002). In the assumptions, the agreement on the basic values of sustainability, as listed and described in GRI (2011), ISO 26000 and UN Global Compact (2011), among all the parties involved may have been included.

Project procurement should be described in detail in its characteristics of schedule (deadlines, deliverable dates and time), cost, quality, and quantity. Particularly, the sustainability quality requirements should be added to the technical requirements. The description of the supplies should be sufficiently clear and comprehensible, so that the supplier can provide them correctly and in accordance with the agreements for the relevant deliverables. Specifically, what is included in the supplies and what is excluded from them should be described.

Procurement may concern both human and material resources. Suppliers should be chosen sustainably, in the respect for the personal dignity, and developing a careful cost control. The necessary kind of resource should be described in detail in its qualitative and quantitative characteristics. It should be indicated when the resource will be needed and for how long.

In addition to the traditional criteria adopted in the selection of the suppliers, sustainability requirements should be added and the vendors should be selected performing a check of their references and considering some weights, as shown for instance in Table 5.6, relevant to a single supply X. The evaluation of each supplier refers to the offer presented and is expressed through a score ranging from 1 (poor) to 10 (excellent), and the sum of the relative weights is set equal to 100.

At the end of each supply it is necessary to formally close the contracts signed with the vendors, verifying that the work and the products, services, or results delivered correspond to the technical specifications and to the quality requirements, and that they are acceptable. The compliance with the contract requirements is subject to the control of many stakeholders.

While the project is in progress and at its end, the respect for the sustainability objectives identified for procurement in the initiating stage of the project should always be verified.

Table 5.6 – Selection of suppliers

Supply X	Assessment supplier 1	Assessment supplier 2	Assessment supplier 3	Assessment supplier 4
Cost of the supply (weight 15)	8	6	7	8
Reputation of the supplier (weight 20)	7	8	6	7
Experience in similar supplies (weight 10)	6	6	7	4
Knowledge of the local reality (weight 5)	7	6	7	6
Respect for time (weight 20)	7	7	7	6
Environmental sustainability (weight 10)	6	8	9	7
Social sustainability (weight 12)	6	7	8	6
Economic sustainability (weight 8)	6	6	7	7
Total assessment	675	692	712	648

5.11 Stakeholders

Sustainability is participation.

Project stakeholders, their needs and their demands should be identified and described. Stakeholders may vary with the progress of the project over time. Stakeholders are also future generations. Needs and expectations of stakeholders should be evaluated in the environmental, social and economic area, in the short, medium and long term, locally, regionally and globally.

The introduction of project stakeholder management, as a separate knowledge area, in the fifth edition of the PMBOK® (*Project Management Body of Knowledge*) of the Project Management Institute (PMI®, 2013) and in the new standard 2013 ISO 21500 (*Guidance on Project Management*), is a further step forward towards the introduction at international level of the concept of sustainability in project management.

The addition of the new knowledge area is in line with a growing body of research showing that the involvement of stakeholders is a major factor in the overall success of a project. Further details about some of these researches developed at the University of Vienna (Austria) can be found in the references at the end of the book.

Stakeholders can be the project sponsors, the project customers, the senior management, the company departments, the members of the project team, the consumers, the users, the suppliers, the consultants, the business partners, the institutions, and any other interested parties. They may influence the project, can be affected by it, or may consider themselves affected by decisions, activities, results or outcomes of the project itself.

For the identification of the project stakeholders and of their mutual relations and dynamics, in addition to the traditional tools and techniques as for instance the so-called "brainstorming", the systemic constellations can be used, because a project may be considered as a social system (Gareis et al., 2013). See also section 5.9.

Formerly, the purpose of project stakeholder management was solely to manage their expectations, distributing the relevant, necessary and required information to them. Their management fell in the knowledge area of project communications. Now it means instead involving stakeholders actively and effectively, from initiating, in making decisions about planning, executing and controlling the project itself.

Considering the active participation of stakeholders in the development of projects is necessary. Just think to the consumer demands in the environmental field, which then affect the production of goods (as it happens for instance in the market of hybrid cars). This activism can show both in favor and against a project, requiring in any case to be managed. The management becomes critical in today's reality, where data, information, communications and actions are intertwined and develop with a rapidity never experienced before.

Sustainability is therefore also the transition from the management *of* stakeholders to the management *for* stakeholders. And this goal can be achieved more accurately and more easily if adequate structured project management processes and procedures are available.

Considering the definition of sustainable development of the Brundtland Report, also the future generations could be added to the stakeholder register. They are a third party with high interest and legitimacy to be represented, but low power, influence and impact at the present time. Future generations cannot be directly involved in decision-making and in the execution of the project, but their needs can be taken into account. It should be also measured the degree of satisfaction of future generations in regard to the project, in the same way in which it is usually done for each key objective of the project. This inclusion could add commercial value to projects, with tangible and intangible benefits in the medium and long term. The following questions should tentatively be asked. Does the project, or its management, or its results, products and services, damage or destroy something of value that can not be rebuilt? Will the project manager be satisfied about this project, or about its management, or about its results, products and services also in the future, looking back at the work done? Does the project, or its management, or its results, products and services provide conditions of enduring well-being to society?

The management of project stakeholders concerns also the management of the project team. And managing a work group sustainably means creating a comfortable working environment, inspiring positive goals, emotions and relationships.

The project manager should develop an appropriate stakeholder sustainable management plan. Well planned projects should describe and include negotiations with all reasonable stakeholders, eager to share their best knowledge on the topics of interest for the benefit of the project itself.

While the project is in progress and at its end, the respect for the sustainability objectives identified for stakeholders in the initiating stage of the project should always be verified.

5.12 Verification of the project results

Also the benefits of the project should be identified and assessed in terms of global sustainability. Not only the economic results of the project should be verified, but also the environmental and social ones in the short, medium and long term, locally, regionally and globally.

In the project charter the criteria for assessing the success of the project, that is the measurable parameters which will show whether the project objectives have been achieved satisfactorily, should have been previously outlined.

The results of the project can be verified while progressing over time according to the planned milestones or deliverables, controlling for each deliverable or milestone if the estimated delivery time described in the schedule, the planned quality parameters, the estimated costs, and the sustainability parameters have been met. Variances should be measured, both positive and negative, with respect to what has been planned, the reasons for such deviations should be provided, and any possible actions executed for the compliance with what has been planned should be described.

With reference to a single deliverable or milestone X it is possible, for instance, to perform the assessments listed in Table 5.7, where S indicates the achievement of the planned objectives, N denotes the non achievement of the planned objectives, a variance + indicates a positive deviation for the project (an improvement, an added value), and a variance - indicates a negative deviation for the project (a deterioration, a delay, an economic loss).

The variances, both positive and negative, will be considered acceptable if they are included in the range of values previously indicated for each item by the criteria written in the project charter and in the project management plan for the assessment of the success of the project.

Table 5.7 – Attained objectives by milestone or deliverable

Deliverable/Milestone X	S/N	Variance +/-	Cause	Actions
Respect for time				
Respect for quality				
Respect for cost				
Respect for environmental requirements				
Respect for social requirements				
Respect for economic requirements				

In the previous Table 5.2 it was highlighted how the general objectives of the project could be referred to the different knowledge areas.

Comparing the results of the project with the parameters previously listed in Table 5.2, it will be possible to develop for instance Table 5.8, referred to the objectives achieved for each knowledge area, where risk + indicates an opportunity and risk - means a negative event for the project, S indicates the achievement of the planned objectives, N denotes the non achievement of the planned objectives, a variance + indicates a positive deviation for the project (an improvement, an added value) and a variance – indicates a negative deviation of the project (a deterioration, a delay, an economic loss).

Each objective has also its own geographical and time allocation. Also here the variances, both positive and negative, will be considered acceptable if they are included in the range of values previously indicated for each item by the criteria written in the project charter and in the project management plan for the assessment of the success of the project.

Table 5.8 – Attained objectives by knowledge areas

Knowledge area	Attained economic objectives (S/N)	Variance from the objective (+/-)	Attained environmental objectives (S/N)	Variance from the objective (+/-)	Attained social objectives (S/N)	Variance from the objective (+/-)
Scope						
Time						
Cost						
Quality						
Human resource						
Communications						
Risk +/-						
Procurement						
Stakeholder						

Tables 5.7 and 5.8 can be used in the processes of project monitoring and controlling or to provide periodical information to stakeholders on the progress of the project. They could also be useful in the closing stage of the project, to summarize its entire proceeding and to deduce what has been learnt through the project itself (lessons learned).

The completed project should leave the positive feeling of a well done work.

6. Conclusions

6.1 Conclusions

The best way to manage a project sustainably is:

- Following a standard;
- Including sustainability objectives from initiating.

For the success of the project it is necessary that the sustainability objectives have already been included in the documents, that initiate the project itself. This means that the sustainability objectives have already been considered in the analysis of the investment and that the project manager and her or his project team have been chosen according to their specific expertise in the field and know and agree on its core values, agreed also by the company itself and included in the business strategies.

Projects, by their nature, are temporary and have a well-defined beginning and end; their products, services, or results instead last over time.

For sustainable projects the investment analysis does not develop any more only according to the classical schemes:

- Economic assessment of the investment (the investment pays back mainly with the cash flow of the project);
- Cost-benefit assessment (study of the investment, developed in such a manner and detail to satisfy the needs of the investors);

but includes also:

- Social cost-benefit assessment (considering the impacts of the investment in the long term and on all the interested parties);
- Environmental impact analysis (considering the environmental impacts);
- Analysis of the context of the investment (considering the relations with other correlated investments).

The objectives of the project should include environmental, social and economic sustainability objectives, in the short, medium and long term, on a local, regional and global extent, and should be specific, measurable, agreed, realistic and allocated in the time.

Among the principles of sustainability there is also to consume the income and not the capital. That is, all the resources of the project, also the human ones, should be consumed according to ways and time that would allow them to regenerate.

The project manager could have low influence on the organizational structure, on the choice of raw materials and energy and, therefore, on the production cycle itself, as well as on the available equipment, on the economic resources, and on the selection of suppliers.

The project manager instead affects directly and daily the organization of the work and the management of resources, particularly the human ones, and it is precisely in these areas that this handbook can find wide application.

The sustainable management of a project implies an ethic and enlightened organization of the work, that protects human resources and that performs through personal daily actions. The sustainable project manager knows how to take care of all her or his environmental, social and economic resources, and knows how to maintain their efficiency over time.

The project team can be managed sustainably, valuing its strengths and managing positive goals, positive emotions, and positive relationships.

In the assessment of risks and opportunities, they should be considered both risks and opportunities for the project and risks and opportunities for stakeholders in the environmental, social and economic areas, in the short, medium and long term, on a local, regional and global extent, basing upon agreed values.

In projects that create sustainable products, results or services, sustainability can be understood as a quality requirement and should be included in the technical specifications of the product, service, or result provided.

The selection of the suppliers should be guided by the respect for the sustainability objectives of the project. Also suppliers, contractors and subcontractors should agree on the core values of sustainability.

Special attention and caution should be given to the management of project stakeholders. The management *of* stakeholders turns into the management *for* stakeholders. All the parties directly or indirectly interested in or affected by the project are actively and effectively involved, from initiating, in making decisions on planning, executing and controlling the project itself. In order to avoid manipulation or unpleasant situations, for instance with dishonest stakeholders, the greatest caution is needed in this new approach, which always needs the common agreement on the basic values of the sustainable development.

Progressing the project over time, at the intermediate milestones and at its completion, the achievement of the planned sustainability goals should be verified and all that has been learnt from the particular project should be analyzed.

For specific environmental, social or economic knowledge the collaboration of experts in the field could be required.

Only some projects are sustainable by nature, but all projects can be managed sustainably.

Glossary

Glossary of terms used in sustainability

Concise glossary of terms used in environmental, social, and economic sustainability.

Acid rains

They may have a man-made origin: combustion of coal, SO_2 (sulfur dioxide), chemical plants, waste water treatment plants, biological oxidation, H_2S (hydrogen sulfide).

They may have a natural origin: volcanic activity, biological processes.

Main effects:

- − Damage to plants;
- − Damage to aquatic life;
- − Corrosion of ferrous materials;
- − Deterioration of stone materials.

Aerobic

Organisms that use O_2 as an oxidizing agent.

Aerobic transformation

An aerobic transformation is a complete oxidization of the organic matter through the action of bacteria and leads to the formation of: CO_2, H_2O, NO_2, NO_3. The mineralization of carbohydrates and of fats leads to the formation of CO_2 and H_2O. The mineralization of the proteins leads to the formation of ammonia (NH_3), further transformed into nitrite (NO_2) and then into nitrate (NO_3).

Aerosol

Dispersion of fine solid or liquid particles in a gas. They can be in form of:

- powder (solid particles from crushing);
- exhalation (solid particles from vapor condensation);
- smoke (solid and liquid particles from combustion);
- fog (liquid particles from condensation);
- smog, that is a mixture of smoke and fog (liquid particles from condensation plus solid and liquid particles from combustion).

Air pollution

It can be caused by: acid rain, volcanic eruptions, traffic, heating plants, industrial plants, power plants, air conditioning plants, handled powdery material, and paints.

Air pollutants may be in the form of gas (substances in the gaseous phase), vapor (solutions of liquids in air), or aerosol (dispersion of solid or liquid phases in a gas).

The unpolluted air has an average composition as follows:

78,1%	Nitrogen (N_2)
20,9%	Oxygen (O_2)
1,0%	Other substances (Ar, CO_2, Ne, He, CH_4, H_2, CO, Xe, O_3, NO_2, NH_3, SO_2)

Algae

Unicellular or multicellular, photosynthetic, typically autotrophic organisms, without differentiation into tissues. They produce organic matter from CO_2 and H_2O. Gaseous oxygen O_2 is the waste product of their metabolism.

Anaerobic

Organisms that use oxidizing agents different from O_2.

Anaerobic transformation

An anaerobic transformation is a slow and incomplete reduction of the organic substance through the action of bacteria and leads to the formation of: NH_3, CH_4, H_2S, sulfides, volatile acids, and mercaptans.

Autotrophs

Organisms that use CO_2 as a carbon source. Producers that synthesize simple elements into complex substances. Autotrophs are for example plants.

Bacteria

Unicellular microorganisms of the size of microns, with weight of the order of 10^{-6} μg. They may be spherical bacteria (cocci), rod-shaped bacteria (bacilli), curved rod-shaped bacteria (vibrios), or spiral bacteria (spirilla). They feed through their semipermeable cell membrane (osmosis/enzymes). They reproduce by binary fission reaching 250 replications in 24 hours. The bacteria can be:
- saprophytic (feeding on inert, non-living organic matter);
- autotrophic (using CO_2 as a carbon source);
- heterotrophic (using organic substances as a carbon source);
- aerobic (using O_2 as oxidizing agent);
- anaerobic (using oxidizing agents different from O_2).

Bioaccumulation or bioconcentration

Mechanism whereby certain pollutants, such as DDT, PCB, Hg, vinyl chloride, and E. Coli accumulate in living organisms.

Biologic accumulators of pollutants in *water*: bivalve mussels, bottom mullets, crustaceans, small shrimps, and tuna fishes.

Biologic accumulators of pollutants in *soil*: earthworms and plants.

Root absorption: plants absorb elements in ionic form from the soil solution through the root system.

Leaf absorption: leaves can absorb cations, anions and molecules of small size in contact with salt solutions (aerosol and rain water containing elements leached from atmosphere).

Biocenosis
Community consisting of plants and animals living in a particular habitat.

Biocentrism
Opposite of anthropocentrism. Needs and rights of human beings are not more important than those ones of other living beings such as plants and animals.

Biogeochemical cycles
Nutrient cycles from the abiotic world to the biotic world and vice versa.
Examples:
  – Water cycle;
  – Carbon cycle;
  – Nitrogen cycle;
  – Phosphorus cycle;
  – Sulfur cycle;
  – Oxygen cycle.
The water cycle is important for its possible links with water pollution.
The carbon cycle is fundamental for its potential connections with the greenhouse effect.
The nitrogen cycle is important for its possible links with eutrophication and groundwater pollution.
The phosphorus cycle is fundamental for its potential connections with eutrophication and pollution of marine sediments.

Biological indicators of pollution
Presence of the harmful organism and/or of its metabolites, but also abnormal presence of living organisms, as an indirect sign of pollution in progress. Also the contrary, that is the presence of

particular living organisms, which indicate indirectly the absence of pollution.

Presence of pathogenic microorganisms, but also of seagulls or mice (for the presence of waste).

Biological magnification

Mechanism where some toxic substances, such as DDT, PCB, and Hg, concentrate increasingly in living organisms during subsequent steps in the food chain.

Biotic Extended Index

Measure of biodiversity and of the health of rivers. It helps in the development of thematic maps with different colors for different river trunks.

Biotope

Uniform chemical and physical environment.

Combustion

It can have man-made origin: vehicular traffic, heating plants, thermal power plants, industries, incomplete combustion processes, automobile exhaust gases, waste treatment, and waste incinerators.

It may have natural origin: biological processes, lightning, fire, photochemical reactions, breathing, and isoprene (very reactive hydrocarbon generated by plants, which causes characteristic mists).

Main effects:
- Production of toxic primary and secondary pollutants;
- Production of free radicals, which take part in photochemical reactions induced by sunlight;
- Toxic CO (formation of carboxyhemoglobin at the expense of hemoglobin in blood);
- CO_2 is not toxic but contributes to the creation of the greenhouse effect together with other gases (GHGs, greenhouse gases).

Complementary value

The complementary value is the cost value plus the depreciation.

Composting

Production of soil through natural biodegradation of organic matter both from plants and animals. The residue is converted into water, carbon dioxide, minerals and humus by aerobic decomposing microorganisms.

Corruption

Corruption involves: bribery, fraud, extortion, collusion, conflict of interest, money laundering, embezzlement, misappropriation, pressure on trade, abuse of power, illicit enrichment, concealment of money, and obstruction of justice.

See also: United Nations Convention against Corruption, 2003.

Cost value

The cost value is the sum of the market values of all the inputs needed for the production of the good.

Cradle to Cradle® design C2C®

Methodology developed by Michael Braungart and William McDonough to design new processes, products and services, eliminating the concept of waste, using energy from renewable sources, mainly from the sun, and celebrating the cultural and biological diversity. It aims to restore the continuous, both biological and technological, cycles of nutrients, with positive effects in the long term on company profits, environment and human health.

Declaration on Fundamental Principles and Rights at Work

The Declaration on Fundamental Principles and Rights at Work was adopted by ILO (International Labor Organization) in 1998. ILO is an agency of the United Nations.

It is pro: freedom of association, collective bargaining, and equality at work.

It is against: child labor, forced or compulsory labor, and all kinds of discrimination at work.

Disaster
Earthquake, tsunami, epidemic, volcanic eruption, flood, bradyseism, landslide, inundation, fire, etcetera.

Dissolved oxygen
Amount of dissolved oxygen in water. It decreases while temperature increases. It decreases while concentration of minerals (salt) increases. It increases while pressure increases.. About 10 mg/L at atmospheric pressure and at 20°C in fresh water. About 8 mg/L at atmospheric pressure and at 20°C in seawater.

Dissolved solids
All solid substances dissolved in water, forming part of the filterable solids, which are not retained through the filter or through the centrifuge (also colloidal fraction). (mg/L)

The electrical conductivity measures its fraction formed by mineral salts. (μohm/cm^2)

Ecological factors
   – Orographic: altitude, slope, and orientation;
   – Climate: temperature, light, rainfall, humidity, and wind;
   – Mechanic: water and wind erosion, freeze-thaw cycles;
   – Edaphic: physical-chemical properties of the soil (water content, nutrients, pH, etc.);
   – Biotic: relationships between living organisms;
   – Anthropic: activities of human beings.

Ecological niche
Environmental unit consisting of living species and of their chemical and physical environment.

Ecology

From the Greek *oikos* = home, place of residence. Study of reciprocal interactions between physical environment and living organisms. Branch of biology that studies the relationships between organisms and their environment.

Economic good

The economic good (commodity) is a product or a service, which has characteristics of utility, usability and limited availability. It can be durable or nondurable, present or future.

Economic value distributed

It indicates how the project creates wealth for its stakeholders and includes:
- Payments to employees (wages, training courses, refresher courses, seminars, etc.);
- Payments to contractors;
- Payments to suppliers;
- Donations or services voluntarily offered to the community;
- Direct investments in infrastructures for the community;
- Payments to financial investors (for instance shareholders, loans);
- Payments to the local authorities and governments (through taxation).

Economy

Efficient and rational use of resources in the production of goods and services.

(*... the economic analysis of an investment differs from the financial analysis being the former referred to the whole community and the latter to the private operator, who undertakes it ...*) (Forte, 1977).

Ecosystem

Set of interactions among biotic communities, living organisms, and the surrounding chemical and physical environment. Human beings are a biotic component of our ecosystem.

Energy from non-renewable sources

Energy derived from natural processes that develop during geological eras. Energy from non-renewable sources: coal, natural gas, oil and its by-products, and nuclear.

Energy from renewable sources

Energy derived from natural processes in a continuous cycle. Energy from renewable sources: sun, wind, tides, hydropower, biomass, geothermal resources, (non for food) biofuels, hydrogen, and ethanol.

Environment

What is around, surrounding. Set of the external conditions. Space (place and time) where we live.

Eutrophication

Hypoxia in a body of stagnant water for the excessive presence of nutrients (phosphorus and nitrogen), which leads to an abnormal growth and to the subsequent death of phytomass (algal bloom). In this way some processes of decomposition start at the bottom, resulting in decreased dissolved oxygen and in the death of fishes and of other aquatic organisms.

It involves an abnormal development of algae and a water coloring (green, brown, or blue).

It also increases the amount of dead algae that settle at the bottom in thick layers, starting anaerobic transformations (fermentation or putrefaction) and the formation of NH_3, CH_4, H_2S, sulfides, volatile acids, and mercaptans. It causes a reduction in dissolved oxygen and generates unpleasant odors. The reduction in the content of O_2 compromises the survival of many aquatic species (fish murrain).

It is caused by the discharge in the water of nitrogen and phosphorus compounds. Also detergents (or surfactants) can contain phosphates. These compounds are actually fertilizers and therefore are nutrients for algae.

This phenomenon affects particularly the lakes, but also the seas with poor water exchange. It is increased by shallow waters, by poor water exchange, by high temperatures, by urban wastewater discharge, and by agricultural and livestock inflows.

Exchange value or market value

The exchange value or market value is the most probable value, expressed in money, of a commodity traded in a market and it is a historical datum connected with the particular market.

Finance

Management of the money.

(... *the economic analysis of an investment differs from the financial analysis being the former referred to the whole community and the latter to the private operator, who undertakes it* ...) (Forte, 1977).

Food chain or trophic chain

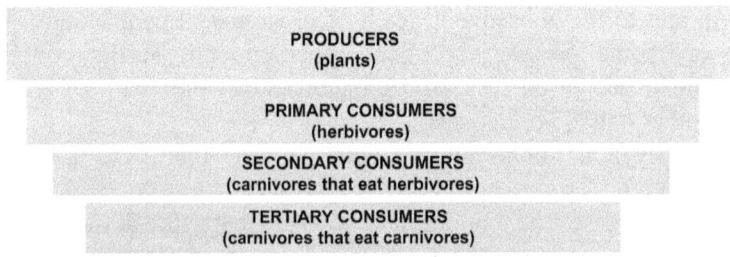

Fungi and mold

They proliferate in a damp environment and generally have a heterotrophic, aerobic, and saprophyte metabolism. They have dimensions of the order of microns and are provided with colored pigments. They can form long filamentous nucleated structures called hyphae. They are indispensable in composting for the transformation of biodegradable organic residues.

Global sustainability

"Global sustainability is the attainment of enduring economic, social, and environmental well-being of all elements of society". (Source PMI® Project Management Global Sustainability Community of Practice, Business plan, 2009).

Greenhouse effect

Carbon dioxide is normally present in the atmosphere in relatively low amounts (<1%). An excess of CO_2 absorbs infrared rays (with high wavelength and low frequency) coming from the earth's surface, creating a barrier to the thermal radiation and a consequent increase in temperature. Carbon dioxide is produced with combustion and removed with plant photosynthesis.

Other greenhouse gases, which are considered among the main cause of climate change, are: methane CH_4, nitrous oxide N_2O, hydrofluorocarbons HFCs, perfluorocarbons PFCs, and sulphur hexafluoride SF_6.

Health

"Health is a state of complete physical, mental and social well-being and not merely the absence of disease or infirmity". Definition of the World Health Organization.

Heterotrophs

Organisms that use organic substances as carbon source. Consumers who are not capable of synthesizing simple elements in complex substances. For instance, animals are heterotrophs.

Nanoparticles

Nano scale particles are so tiny that they cannot be detected by microscopes. A nanometer is 100,000 times thinner than a strand of hair. Nanomaterials can exhibit unique optical, mechanical, magnetic, conductive and sorptive properties different than the same chemical substances in a larger size. (Source: EPA, 2014)

Non-renewable raw materials

Raw materials or goods which are subject to exhaustion: minerals, coal, oil and its by-products, natural gas, and metals. The reserves of these resources regenerate in geological eras.

Non-settleable (colloidal) suspended solids

All solid substances suspended in water that in a 2 hours period do not settle to the bottom of a cone-shaped container (Imhoff cone). (mg/L, p.p.m., g/m^3)

Organic matter

The organic matter consists mainly of carbonaceous compounds (fats and sugars) and nitrogen compounds (proteins and amino acids): C, H, O, N, P, S.

Ozone

O_3

Gas that filters the radiation of low wavelength, that is of high frequency. It forms a barrier to high-energy ultraviolet radiation in the stratosphere, which damage the genetic material of living organisms. Main causes for the reduction of the ozone layer in the atmosphere are: halogens, chlorofluorocarbons CFCs, and hydrochlorofluorocarbons HCFCs. Ozone is a pollutant at the level of the urban atmosphere, because it contributes to the photochemical pollution, conversely in the stratospheric level it filters the ultraviolet radiation of higher energy.

Photochemical pollution

It is characterized by photochemical reactions induced by sunlight. It affects above all eyes and respiratory system. It produces a chain of photochemical reactions, involving unburned hydrocarbons, propellant gases or gases used in refrigeration cycles. Unstable compounds, free radicals, isoprene and halogens are very reactive, especially in the presence of ozone, which at the level of urban atmosphere is a pollutant. Conversely, at the stratospheric level it filters the ultraviolet radiation of higher energy (high frequency and low wavelength).

Pollution

Alteration of the quality and of the characteristics of the natural resource, also as a function of its intended use over time.

Post-disaster reconstruction

Stage following the phases of disaster response and relief. It usually occurs approximately from 4 to 6 months after the actual disaster.

Production

Transformation of natural or material goods into economic assets of greater utility.

Project

Ordered and detailed plan of work to execute something (project).

Set of calculations, drawings, and documents needed to define unequivocally the idea based on which any construction (product or good) can be created (design).

Temporary endeavor undertaken to create a unique product, service, or result (PMI®, 2013).

Remediation

Restoring an altered condition of the environmental components respecting the protection of human health.

Renewable raw materials

Raw materials or goods which are not subject to exhaustion: wind, sun, tides, air, water, soil, and non-food biomass. Because of pollution, air, water and soil may get exhausted for qualitative characteristics. The reserves of these resources regenerate in a short time through biogeochemical cycles.

Saprophytes
Organisms that feed on inert, non-living organic matter.

Settleable suspended solids
All solid substances suspended in water that in a 2 hours period settle to the bottom of a cone-shaped container (Imhoff cone). (mL/L)

Social use value
The social use value (or economic value) of a good refers to the appreciation that the community has of it in terms of its collective usefulness and usability.

Soil
Polyphasic product of the environment formed by chemical substances and biological elements.
Chemical substances: C, H, O, N, P, K, S, Ca, Mg, Fe, Mn, Zn, Cu, B, Mo, H_2O (liquid and vapor), CO_2, O_2, CH_4, H_2S, NH_3.
Biological elements: earthworms, rodents, reptiles, snails, insects, spiders, mites, millipedes, annelids, rotifers, flagellates, ciliates, amoebae, bacteria (nitrifying bacteria, sulfur bacteria), micro-algae, and fungi (yeasts).

Soil pollution
Pollution caused by heavy metals: Cd, Co, Cr, Cu, Hg, Mn, Ni, Pb, Sn, Zn, Mo.
Pollution caused by organic substances: hydrocarbons, pesticides (fungicides, insecticides and herbicides), and PCBs (polychlorinated biphenyls) (chlorinated aromatic hydrocarbons).
It can affect leaf or root absorption. It can also concern the soil formation, the erosion by water or wind, and the desertification.

Suspended solids
All visible solid substances, cause of the turbidity of water, that are retained by the filter or by the centrifuge and are weighed, after

that all the water evaporates with a drying process at 105°C. (mg/L, p.p.m., g/m^3)

To sustain

To help, to protect (passive approach). To feed, to give vigor, to give strength (pro-active approach). To take upon oneself a commitment, a responsibility, a moral or material burden.

Sustainability plan

Long term operational plan that, when the project is completed, will allow the users of the product, service, or result provided to manage it sustainably in the economic, social and environmental aspects.

Sustainable development

"Development that meets the needs of the present without compromising the ability of future generations to meet their own needs" (Source: Brundtland Report, World Commission for Environment and Development, 1987).

Thermal inversion

Normally, the vertical thermal gradient leads to a decrease in temperature of about 1°C each 100 meters in elevation.

In the case of thermal inversion, the vertical thermal gradient is positive instead of negative and causes an accumulation of pollutants in the atmosphere, preventing their dispersion.

It can be caused by: sun radiation of the soil (day/night alternation), and atmospheric subsidence (weather conditions).

Total solids

All solid substances, which remain and are weighed after that all the water evaporates with a drying process at 105°C. (mg/L, p.p.m., g/m^3)

(1 mL = 1 cm^3; 1 kg = 1 L)

Transformation value

The transformation value is the market value of a good after the transformation, minus the cost of the works required for the transformation.

UN Global Compact

(UN, 2011) Commitment that can be voluntarily signed by companies, which wish to adhere to ten universal principles in order to ensure that market, trade, technology and finance advance in ways that globally benefit economy and society.

The ten principles are briefly:

1. Support and respect for human rights;
2. Absence of complicity in abuses with respect to human rights;
3. Recognition of freedom of association and collective bargaining in the workplace;
4. Elimination of all forms of forced or compulsory labor;
5. Abolition of child labor;
6. Elimination of discrimination in the workplace;
7. Caution in dealing with environmental issues;
8. Promotion of environmental responsibility;
9. Development and diffusion of technologies that are not harmful to the environment;
10. Opposition against corruption in all its forms, including extortion and bribery.

Unicellular microalgae

Algae of the size of about 10 micron, with colored pigments.

United Nations Convention against Corruption

The United Nations Convention against Corruption was adopted in 2003 and entered into force in 2005.

It is in favor of: democracy, human rights, environmental protection, quality of life, economic development, honesty, transparency, accountability, international cooperation, and sustainable development.

It is against: corruption, market distortion, security threats, poverty, and money laundering.

Universal Declaration of Human Rights
The Universal Declaration of Human Rights was adopted by the General Assembly of the United Nations in 1948.

It is pro: freedom, justice, peace, dignity, freedom of speech and belief, freedom of opinion and expression, social progress, freedom of thought, conscience and religion, economic, social and cultural rights, work, rest, education, health, food, art, and scientific progress.

It is against: tyranny and oppression, all kinds of discriminations, slavery, and persecution.

Value of subrogation
The value of subrogation is the market value of another commodity with the same utility.

Viruses
Unicellular organisms. Intracellular parasites often pathogens. They are not capable of independent metabolism. They have a size of 0.01 to 0.1 microns and are visible in the electronic microscope. They are devoid of water. Viruses are formed only by nuclear material DNA (deoxyribonucleic acid) or RNA (ribonucleic acid). They inject the DNA strand, which carries the genetic information, into the host cell, using its materials and its energy to reproduce and make it "burst". They have an extreme resistance to cold.

Water pollution indicators
- Microbiological indicators: coliforms, streptococci, etc.
- Biological indicators: algae, helminths, fungi, protozoa, trout (positive presence), etc.;
- Chemical indicators: dissolved oxygen, oxidizability (COD), total organic carbon (TOC), hydrogen sulfide, ammonia, phosphates, chlorides, Hg, Cd, Se, Zn, Sn, N, P, nitrates, pesticides, etc.

- Physical indicators: color, turbidity, odor (H_2S), taste, temperature.

Microorganisms: algae, non harmful bacteria, saprophyte fecal bacteria, pathogenic intestinal microorganisms (bacteria of typhoid, paratyphoid, dysentery, cholera, salmonella, hepatitis virus, polio virus, enterovirus), and eggs of intestinal worms (tapeworms, roundworms).

Water pollution can be: from sewer, agricultural, livestock, industrial, industrial nuclear. It can also be caused by: maritime, river and lake transportation, and off-shore oil extraction.

Xenobiotics

Synthetic substances, which do not have the corresponding decomposing organism in the natural environment.

References

Barnard L.T., Ackles B., Haner J.L. (2011), *Making Sense of Sustainability Project Management*, Explorus Group Inc., Canada.

Boulding K.E. (1966), Essay: *The economics of the coming spaceship earth*, USA.

Braungart M., McDonough W. (2002), *Cradle to cradle: remaking the way we make things*, North Point Press, USA.

Carboni J, Gonzalez M., Hodgkinson J. (2013), *The GPM® Reference Guide to Sustainability in Project Management*, GPM Global, USA.

Deshpande A. (2011), *Equator Principles: Do they make business sense?*, Eco Business, USA, January.

EPA., U.S. Environmental Protection Agency (2010), *Ozone-depleting Substances*, USA.

EPA, U.S. Environmental Protection Agency (2011), *Greenhouse Gas Emissions*, USA.

EPA, U.S. Environmental Protection Agency (2014), *Questions about nanotechnology*, USA.

ESI - Environmental Sustainability Index (2005), *2005 ESI Report*, World Economic Forum's Annual Meeting, Davos, Switzerland.

Esposito C. (2007), *Il Cimitero delle Fontanelle*, Oxiana, Naples.

Forte C. (1977), *Valore di scambio e valore d'uso sociale dei beni culturali immobiliari*, Formez, Naples.

Freeman H. M. (1988), *Standard Handbook of Hazardous Waste Treatment and Disposal*, Mc Graw-Hill Book Company, USA.

Gareis R. (2012), WU Vienna, Austria, & RGC, *Rethinking project management*, PMI® EMEA ROWS Marseilles, May.

Gareis R., Huemann M., Martinuzzi A. (2010), *Relating Sustainable Development and Project Management: A Conceptual Model*, PMI® (Project Management Institute), USA.

Gareis R., Huemann M., Martinuzzi A., Weninger C., Sedlacko M. (2013), *Project Management and Sustainable Development Principles*, PMI®, USA.

GRI Global Reporting Initiative (2011), *Sustainability Reporting Guidelines*, GRI, Amsterdam.

Hornby A. S. (1987), *Oxford advanced learner's dictionary of current English*, Oxford University Press, Great Britain.

Huemann M. (2012), WU Vienna, Austria, *Project Stakeholder Management & SD*, PMI® EMEA ROWS Marseilles, May.

IFC (International Finance Corporation) (2006), *Equator Principles*, Standard 2006, E.P. Association, United Kingdom.

ILO (International Labour Organization) (1998), *Declaration on the Fundamental Principles and Rights at Work*, ILO, USA.

ISO (International Organization for Standardization) (2004), *ISO 14000 - Environmental management*, ISO, Switzerland.

ISO (International Organization for Standardization) (2008), *ISO 9000 - Quality management*, ISO, Switzerland.

ISO (International Organization for Standardization) (2010), *ISO 26000 -Guidance on social responsibility*, ISO, Switzerland.

ISO (International Organization for Standardization) (2011), *ISO 50001 - Energy management*, ISO, Switzerland.

ISO (International Organization for Standardization) (2013), *ISO 21500 -Guidance on project management*, ISO, Switzerland.

Mahan B. H. (1979), *Chimica generale ed inorganica*, Casa Editrice Ambrosiana, Milan.

Maltzman R., Shirley D. (2010), *Green Project Management*, CRC Press, USA.

Melisurgo G. (1997), *Napoli sotterranea*, Edizioni Scientifiche Italiane, Naples.

Mendia L. (1959), *Sul controllo della qualità delle acque*, Seminario sull'approvvigionamento idrico delle popolazioni, World Health Organization, Amalfi.

Metcalf & Eddy (1991), *Wastewater Engineering – Treatment, Disposal and Reuse*, McGraw-Hill International Editions, U.S.A.

Ministero dell'ambiente e della tutela del territorio (2006), Direzione generale per la ricerca ambientale e lo sviluppo, *Mille progetti per lo sviluppo sostenibile*, Rapporto di sintesi, Italy, Attività 2000-2006.

Montalenti G., Giacomini V. (1974), *Biologia*, Sansoni, Florence.

Morgese P. (1993), *Criteri di qualità delle acque destinate a scopo potabile*, Tesina di Ingegneria sanitaria, Università degli Studi di Napoli "Federico II", Facoltà di Ingegneria, Scuola di specializzazione in ingegneria sanitaria ed ambientale, A.A. 1992/1993.

Morgese P. (1994), Tesi di diploma di specializzazione: "*Trattamento e smaltimento di residui industriali siderurgici in stabilimenti dismessi – Aspetti normativi e criteri per la bonifica dei materiali, del suolo e delle falde*", Università degli Studi di Napoli "Federico II", Facoltà di Ingegneria, Scuola di specializzazione in ingegneria sanitaria ed ambientale, A.A. 1993/1994.

Morgese P. (1997), *Collaudo tecnico delle operazioni di bonifica nello stabilimento ILVA di Bagnoli*, Simposio internazionale di ingegneria sanitaria ambientale, Ravello, Villa Rufolo, 3 - 7 June.

Morgese P. (2007), Lezioni di *Interventi antropici ed elementi dello sviluppo sostenibile*, Progetto IFTS PON Tecnico superiore per il monitoraggio e la gestione del territorio e dell'ambiente, IPSAR Cicciano Naples.

Morgese P. (2009a), Lezioni di *Bonifica dei siti contaminati: trattamenti chimici*, Master in Ingegneria sanitaria ed ambientale: ciclo dei rifiuti e bonifica dei siti contaminati, Facoltà di Ingegneria Università di Napoli "Federico II", A.A. 2008/2009.

Morgese P. (2009b), *Il project management nelle bonifiche dei siti contaminati*, Seminar, Facoltà di Ingegneria, Università di Napoli "Federico II", 23 November.

Morgese P. (2010), *La sostenibilità ambientale e il ruolo dell'ingegnere*, Notiziario dell'Ordine degli Ingegneri della Provincia di Napoli, Numero 4, November-December.

Morgese P. (2011a), *Integrating global sustainability into project management: the human resource knowledge area*, PMI® (Project Management Institute) Project Management Global Sustainability Community of Practice, USA.

Morgese P. (2011b), *Un nuovo legame tra energie rinnovabili e project management: gli "Equator Principles"*, Notiziario dell'Ordine degli Ingegneri della Provincia di Napoli, Numero 2.

Morgese P. (2011c), *La valutazione numerica della sostenibilità ambientale di un'azienda*, Notiziario dell'Ordine degli Ingegneri della Provincia di Napoli, Numero 5, September-October.

Morgese P. (2012a), *La sostenibilità e il project management – Progetti sostenibili*, Seminario Facoltà di Ingegneria Università di Napoli "Federico II", 27 January.

Morgese P. (2012b), *Sul valore d'uso sociale dei beni ambientali*, Notiziario dell'Ordine degli Ingegneri della Provincia di Napoli, Numero 2.

Morgese P. (2013a), *La gestione degli stakeholder nei progetti: un ulteriore passo verso la sostenibilità*, Notiziario dell'Ordine degli Ingegneri della Provincia di Napoli, Numero 1.

Morgese P. (2013b), *Projects for post disaster reconstruction are sustainable by nature but, how to manage them in sustainable ways?*, PMI® Global Sustainability Community of Practice, USA, May.

Morgese P. (2013c), *La gestione dei progetti nella ricostruzione post disastro*, Seminar, Facoltà di Ingegneria Università di Napoli "Federico II", 24 May.

Morgese P. (2013d), *La gestione dei progetti di ricostruzione post disastro*, Notiziario dell'Ordine degli Ingegneri della Provincia di Napoli, Numero 3.

Orefice M. (1984), *Estimo*, UTET, Turin.

Perry R. H. et al. (1995), *Perry's chemical engineers' handbook*, McGraw-Hill Book Company, New York.

Nyangon J. (2011), *Rebalancing the economics of greening*, PMI® Global Sustainability Community of Practice, USA.

PMI® (2010), *Code of Ethics and Professional Conduct*, Project Management Institute, USA.

PMI® (2012), *The PMI ethical decision-making framework*, Project Management Institute, USA.

PMI® (2013), *PMBOK® Guide – Fifth Edition*, Project Management Institute, USA.

PMI®, PMIEF® (PMI® Educational Foundation) (2013), *Project Management Methodology for Post Disaster Reconstruction*, Italian translator Paola Morgese, Project Management Institute, USA.

PM NETWORK® (2009), *Rising to the occasion*, Project Management Institute, USA, December.

PM NETWORK® (2011), *The buzz – Power spike*, Project Management Institute, USA, March.

Ragazzini G. (1989), *Dizionario Inglese/Italiano Italiano/Inglese*, Zanichelli, Bologna.

SAI Social Accountability International (2008), *SAI standard 8000 –* Social Accountability International, New York.

Sequi P. (1991), *Chimica del suolo*, Patron Editore, Bologna.

Silvius G. A. J., Schipper R., Planko J., van den Brink J., Köhler A. (2012), *Sustainability in Project Management*, Gower, United Kingdom.

UN United Nations (1948), *Dichiarazione universale dei diritti umani*, United Nations, New York.

UN United Nations (2000), *Sphere Handbook*, United Nations, New York.

UN United Nations (2003), *United Nations Convention Against Corruption*, United Nations, New York.

UN United Nations (2011), *Global compact*, United Nations, New York.

Vienna Hospital Association (2010), *Vienna North Hospital's Charter on Sustainability*, Vienna Hospital Association, Vienna.

Vismara R. (1992), *Ecologia applicata*, Hoepli, Milan.

WCED (1987), World Commission for Environment and Development, *Brundtland Report*, United Nations, New York.

Zingarelli N. (1988), *Vocabolario della lingua italiana*, Zanichelli, Bologna.

Acknowledgments

The author acknowledge her gratitude to the never forgotten university professor Luigi Mendia for his teachings; to the friend and class-mate Tina D'Ambra for kindly proofreading the English version; to the colleague Joseph Nyangon for the peer review and the precious suggestions; to Gennaro Cimmino for the photographic portrait; to Leda, Filuccia and Felis for their company while writing.

Index

Notes

Notes

www.ingramcontent.com/pod-product-compliance
Lightning Source LLC
Chambersburg PA
CBHW051719170526
45167CB00002B/720